Springer Climate

Series Editor

John Dodson, Institute of Earth Environment, Chinese Academy of Sciences, Xian, China

Springer Climate is an interdisciplinary book series dedicated to climate research. This includes climatology, climate change impacts, climate change management, climate change policy, regional climate studies, climate monitoring and modeling, palaeoclimatology etc. The series publishes high quality research for scientists, researchers, students and policy makers. An author/editor questionnaire, instructions for authors and a book proposal form can be obtained from the Publishing Editor.

Now indexed in Scopus® !

Annapurna Boruah

CO_2 Geosequestration: Capturing Carbon for a Sustainable Future

Carbon Dioxide Storage in Geological Media

Springer

Annapurna Boruah
University of Petroleum and Energy Studies
(UPES)
Dehradun, Uttarakhand, India

ISSN 2352-0698 ISSN 2352-0701 (electronic)
Springer Climate
ISBN 978-3-031-81023-7 ISBN 978-3-031-81021-3 (eBook)
https://doi.org/10.1007/978-3-031-81021-3

This Springer imprint is published by the registered company Springer Nature Switzerland AG
The registered company address is: Gewerbestrasse 11, 6330 Cham, Switzerland

Preamble

As the global community is struggling with the effects of climate change, reducing atmospheric carbon dioxide (CO_2) levels has become a critical task. CO_2 geosequestration—capturing CO_2 emissions and tightly storing them underground—deals an encouraging solution to mitigate these waves. This book discusses the science, technology, and application of CO_2 geosequestration across different geological formations, providing a wide-ranging understanding of its possibility in the fight against climate change.

Chapter 1 “Climate change and CO_2 Geosequestration”, sets the step by inspecting the importance of CO_2 capture and storage in decreasing greenhouse gas emissions. It explains the fundamental role of geosequestration in transitioning towards a low-carbon future and outlines the technical and environmental considerations.

Chapter 2 “Reservoir Screening Criteria for CO_2 Storage” delves into the essential geological factors that estimate the suitability of subsurface formations for CO_2 storage. It covers key characteristics such as porosity, permeability, and structural integrity that make particular reservoirs/rocks ideal for long-term sequestration.

Moving into specific formations, Chap. 3 discusses the CO_2Sequestration in Shale, and the potential of organic-rich shales for trapping CO_2. The shales, with their fine-grained structure, offer a unique medium for CO_2 storage; however, shales also show challenges related to pressure buildup and its tight nature.

Chapter 4 explains how the CO_2Fracturing in Shale can shift to the techniques used to enhance CO_2 injection in shale formations. Hydraulic fracturing, or “fracking,” combined with CO_2, presents an innovative approach for enhancing the permeability of these tight reservoirs, enabling more effective storage.

Chapter 5 regarding CO_2-Enhanced CBM Recovery from Coal explores how CO_2 injection can simultaneously sequester carbon while enhancing coalbed methane (CBM) recovery. These dual-benefit techniques have the potential to enhance energy production along with the addressing environmental concerns.

Chapter 6 explains CO_2 Sequestration in Saline Aquifers as one of the most promising and widely studied geological options for CO_2 storage. Saline aquifers, with their enormous capacity and geographic abundance, are explored in terms of injection strategies, long-term containment, and potential risks.

Finally, Chap. 7 on CO_2 Storage in Basalt and Mineralization delves into the prospects of basalt formations for permanent CO_2 storage through mineralization. This process alters CO_2 into stable carbonate minerals and offers a secure solution to carbon storage for longer period of time compared to other geological Medias.

Through these chapters, the book provides an in-depth interpretation of diverse geological formations and innovative techniques for CO_2 geosequestration. By highlighting both the opportunities and challenges of each approach, it serves as a valuable resource for researchers, geoscientists, engineers, industry professionals, policymakers, and stakeholders working towards a sustainable, carbon-neutral future.

Contents

Chapter 1
Climate Change and CO_2 Geosequestration

Abstract As outlined in the Paris Agreement, achieving net-zero and negative greenhouse gas (GHG) emissions by 2080 to limit global temperature rise to 2 °C, is crucial. Still the fossil fuels are contribution approximately 83% of the world's energy supply. Carbon Capture and Storage (CCS) is considered as a key technology to reduce the growing carbon footprint. CO_2 geosequestration, involves storing CO_2 in the geological media and restrict its release into the atmosphere. The feasible options of geosequestration include underground geological formations, oceans, terrestrial ecosystems, and bio-sequestration. Combining the geological sequestration and as renewable energy exploration can facilitate the carbon emissions reduction. This chapter discusses the CCUS (Carbon Capture, Utilization, and Storage) as a key element in national and global climate strategies, and demonstrating its role in achieving net zero.

Keywords CO_2 capturing technologies · Climate change · CO_2 Geosequestration · Storage · Greenhouse gases

1.1 Introduction

Climate change refers to the long-term changes of temperatures and weather patterns driven by several factors including changes in temperature or large volcanic eruptions [1, 2]. Since the 1800s, human activities have become the main cause of climate change. The main greenhouse gases including carbon dioxide and methane [1–4] get released from driving cars powered by gasoline and heating buildings with coal, clearing land and cutting down forests oil, and gas operations etc. Energy production, industry, transportation, buildings, agriculture, and land use are the key sectors that emitting greenhouse gas emissions, exacerbating global warming and climate change [4–6]. In light of the significant challenges that the global community is currently encountering, including the scarcity of conventional energy resources, environmental degradation, and climate change, it is imperative to pursue sustainable and low-carbon energy sources. The Intergovernmental Panel on Climate Change (IPCC)

A. Boruah, *CO_2 Geosequestration: Capturing Carbon for a Sustainable Future*, Springer Climate, https://doi.org/10.1007/978-3-031-81021-3_1

has identified a number of solutions intended to achieve sustained reductions in CO_2 emissions while simultaneously meeting global energy demands in order to minimise the anthropogenic warming to 1.5–2 °C, as set forth in the 2015 Paris Agreement (Global CCS Institute, 2021). Hydrogen (H_2) production and CO_2 Geological Storage (CGS) are two of these options that are thought to be very promising [7–11]. These techniques not only seek to meet the world's expanding energy needs, but also to drastically lower atmospheric CO_2 levels. The main goal of this research is to investigate the viability, difficulties, and possible benefits of geological storage of CO_2 using a thorough, interdisciplinary, and multiscale approach. Global CO_2 emissions are estimated to be between 36 and 37 billion metric tonnes per year as per recent data [10]. China is the world's biggest emitter, with over 28% of all CO_2 emissions. Its high emissions are a result of its large industrial base, reliance on coal for energy, and fast urbanisation. With 15% of the world's CO_2 emissions coming from it, the United States is the second-biggest emitter [1, 10].

Despite recent declines in coal use, the U.S. still has substantial emissions from transportation, industry, and energy production. India is the third-largest emitter, contributing about 7% of global CO_2 emissions [10, 11]. Its growing economy and energy demands, coupled with heavy reliance on coal, drive its emissions. The EU is a significant emitter, contributing around 9% of global CO_2 emissions. However, individual countries within the EU, like Germany, France, and the UK, have also taken steps to reduce emissions through renewable energy and policy measures. Russia contributes about 5% of global CO_2 emissions, with its energy sector (particularly oil and gas) being a major source [12, 13]. Japan is responsible for about 3% of global emissions, with industrial activities and energy production being key contributors. The technical and commercial viability of carbon storage has been well demonstrated at the megatonne scale. To date, more than 300 million metric tons (Mt) of CO_2 have been successfully sequestered in various geological formations worldwide, highlighting a vast CO_2 storage potential that exceeds 12,000 gigatons (Gt). According to the IPCC, achieving an annual storage rate of 2.3 Gt of CO_2 would provide enough storage capacity for approximately 10,000 years. However, to align with the stringent targets outlined by IPCC model pathways, it is essential to store up to 1200 Gt of CO_2 by the year 2100 [7–9, 13–16].

1.2 Energy Mix and CO_2 Emission

The amount of primary energy consumed globally is rising rapidly. Fossil fuels stays essential to their development, which includes 84% of their energy mix [10]. In 2023, fossil fuels represented 56% of the total consumption of energy, witnessing an annual growth rate of 2%, which is nearly twice the average worldwide rate. The Asia Pacific region contributed to 85% of the Global South's demand and 47% of global demand, with China, India, Indonesia, Japan, and South Korea as the main contributors. In 2023, Southern and Central America, together with the Asia Pacific, demonstrated growth rates exceeding the global average; alternatively, total demand in Africa decreased by 0.4%, with electricity consumption staying unchanged. Electricity demand in North America and Europe declined by 1% and 2%, respectively, due to the effects of energy efficiency regulations, the widespread use of energy-efficient lighting, and evolving consumer habits. Significant geographical differences were observed in the correlations between the sizes of the populations in various areas and the amount of energy that was consumed in those regions in the year 2023 [10–12]. In Africa, South Asia, and Southern & Central America, the average energy consumption per capita was 0.3 petajoules (PJ), while in North America, the CIS, and the Middle East, it averaged 1.8 PJ. In North America, this ratio exceeded the global average of 1.1 PJ by nearly twofold. A analogous pattern was observed in average emissions of greenhouse gases per capita, with Africa, South Asia, and Southern & Central America averaged 2 Mt CO_2, compared to the global average of 6.7 Mt CO_2. North America, the CIS, and the Middle East had a collective average of 11.5 Mt CO_2, almost double the global average. Other Asia Pacific was slightly lower than the global average for energy consumption per capita, nevertheless 0.7 Mt CO_2 above the global average for greenhouse gas emissions each person, largely attributable to China, the largest coal consumer and second-largest oil consumer worldwide. Carbon dioxide emissions resulting from fossil fuel combustion represent the main cause of energy-related greenhouse gas emissions, accounting for approximately 87% of the total. The graph provided as Fig. 1.1 illustrates the increase in global greenhouse gas emissions from the mid-eighteenth century to the present, highlighting the influence of human activities on the environment. Prior to the beginning of the Industrial Revolution, carbon dioxide emissions were essentially nonexistent, which is indicative of a world that relied solely on natural and renewable renewable resources. With the onset of the industrial revolution, the consumption of fossil fuels, such as coal, oil, and natural gas, became prevalent, resulting in an important rise in emissions. The rise in emissions had been comparatively small until the mid-twentieth century, characterised by an upsurge in industrial activities with the growth of the world's population and economy. In 1950, globally CO_2 emissions were roughly 6 billion tonnes. More over twenty billion tonnes of emissions were produced by the year 1990, having nearly doubled over the subsequent four decades. The substantial rise can be attributed to multiple factors, including mass industrial development, increased automotive application, and the expansion of the global economy. From 1990 onwards, the rate of rising emissions stayed at a startlingly fast pace. At present, annual CO_2 emissions exceed 35 billion tonnes [10]. This growth indicates a persistent reliance on fossil fuels, increased demand for energy due to growing populations and growth in industry, particularly in nations that are developing. Regardless of a slight acceleration in the rate of increase in recent years, mainly due to increased awareness and different efforts to reduce carbon footprints, global emissions are still far from their peak. This points to a requirement for significant shifts in worldwide energy production and consumption patterns, in addition to the widespread use of effective capture and storage of carbon technologies; alternatively, CO_2 emissions may persist in growing, enhancing the challenge of global climate change.

Figure 1.2 shows a graph illustrating the development of CO_2 emissions by region, emphasising changes in the worldwide distribution of emissions over time. Until

Fig. 1.1 Annual CO_2 emission from fossil fuels and industry. *Source* Global carbon budget 2023

twentieth century, Europe and the United States were the predominant sources of global CO_2 emissions. In 1900, over ninety percent of emissions stemmed from these regions, and by 1950, they represented over 85% of yearly global emissions. Recent decades have experienced an important shift. The last half of the twentieth century noticed an important rise in emissions globally, especially in Asia, in which China grew an important contributor. Together, the United States and Europe are responsible for less than a third of the world's emissions at this time [10].

Many countries in the Middle East, including as Qatar, the United Arab Emirates, Bahrain, and Kuwait, are among the countries that generate the most carbon dioxide per capita. They are primarily large oil-producing countries with comparatively small populations. The lesser population sizes of these nations contribute to the reality that their total annual emissions are comparatively low, despite the fact that their per capita emissions are quite significant. On the other hand, nations with a larger population, such as the United States of America, Australia, and Canada, have high emissions per capita and, as a result, make substantial contributions to the overall worldwide emissions. These nations have emissions that are nearly three times higher than the average for the entire globe. Due to the reality that there is a significant correlation between levels of prosperity and greenhouse gas emissions per capita, countries that have higher norms of living usually leave higher carbon footprints compared to nations with a lower standard of living. However, even among countries that have comparable standards of living, there are significant distinctions. For instance, in comparison to the United States of America, Canada, or Australia, several European countries have far lower levels of emissions. Portugal, France, and the United Kingdom are just a few of the European nations that have emissions that are relatively near to the average for the entire world. While its neighbours, such as Germany, the Netherlands, or

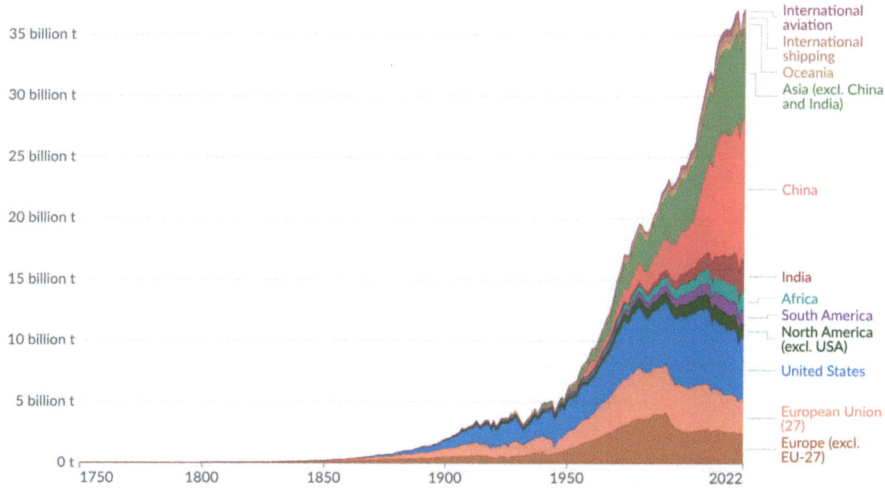

Fig. 1.2 Annual CO_2 emission (fossil fuel and industries) by world region. *Data source* Global Budget 2023

Belgium, have larger emissions, this stands in stark contrast to those countries. A significant percentage of the power that generates in the United Kingdom, Portugal, and France comes from nuclear and renewable sources, but around fifty percent of the electricity that is produced in Germany originates from fossil fuels. The choice of sources of energy is a crucial factor. Although the fact that economic growth is the primary driver of carbon dioxide emissions, results are highly impacted by policy decisions and technology choices. A great number of nations, particularly those located in Sub-Saharan Africa, continue to have extremely low levels of CO_2 emissions per person. An average carbon footprint of roughly 0.1 tonnes per year is found in some of the world's poorest countries, such as Chad, Niger, and the Central African Republic. This quantity is nearly 150 times smaller than the carbon footprints of the United States of America, Australia, and Canada. The average person in the United States or Australia puts out the same amount of pollution in less than two days as the average person in Mali or Niger does in a whole year. Figure 1.3 demonstrates how CO_2 emissions altered globe by sector from 1990 to 2020. The primary driver of emissions remained energy output. This was mostly caused by the combustion of fossil fuels for the production of electricity and heat, which made up around 73% of the total world emissions by the year 2020. As of the year 2023, the principal contributors to greenhouse gas emissions around the world were reflected in the distribution of global CO_2 emissions across important industries. With power and heat generation making up around 31% of total emission and transport contributing 15%, the energy sector remained to be the largest contributor, contributing almost 73% of the total emissions. These sectors are among the most important goals for emission reductions as they are mainly dependent on fossil fuels. Cement and chemical manufacturing are two examples of manufacturing processes that contributed approximately 5.2%

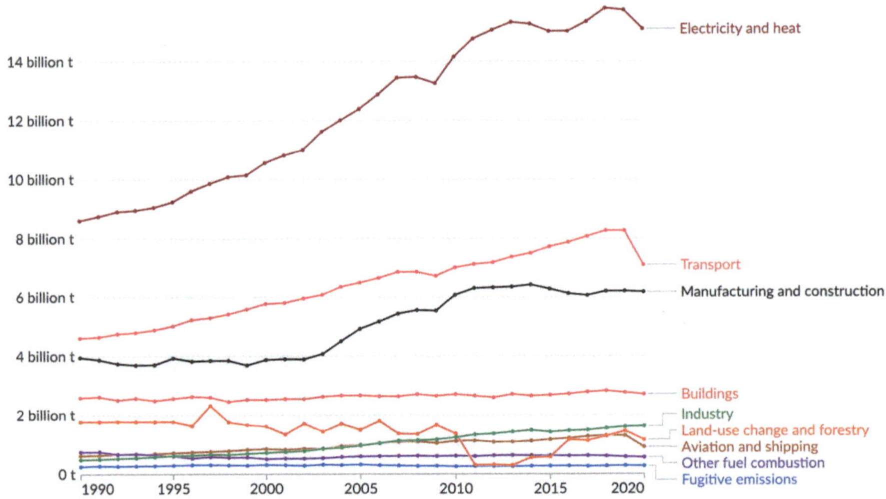

Fig. 1.3 Distribution of CO_2 emission worldwide from 1990 to 2020, by sector. European Commission; EDGAR/JRC; Crippa et al. (2024); IEA and Statista, edgar.jrc.ec.europa.eu

to the total amount of carbon dioxide emissions worldwide. In addition to being the result of the consumption of energy, these emissions are also the result of the chemical reactions that are engaged in the production process. It is projected that around 18.4% of the total emissions originated by the agriculture, forestry, and land use sector. This industry comprises emission from livestock, deforestation, and practices associated with soil management. Finally, the processes of waste management, such as wastewater treatment and landfills, accounted for 3.2% of the total emissions globally (Fig. 1.4).

This distribution highlights the importance of targeting energy production, transportation, and industrial processes in global efforts to reduce CO_2 emissions. Carbon capture, energy efficiency, and transitioning to renewable energy sources are key strategies to mitigate emissions from these sector. Given the scale of emissions from these sectors, Carbon Capture and Storage (CCS) becomes a critical technology in reducing atmospheric CO_2. CCS involves capturing CO_2 emissions at their source—particularly from industrial and energy-related activities—and securely storing it in geological formations, such as deep saline aquifers or depleted oil and gas fields. This process is especially important for sectors that are difficult to decarbonize, like cement and steel production, where emissions arise not only from energy use but also from chemical reactions. Additionally, biosequestration**,** which includes afforestation and soil carbon capture, plays a role in managing emissions from land use and agriculture. The integration of CO_2 sequestration alongside renewable energy adoption and energy efficiency measures is crucial for achieving global climate targets and limiting the impacts of climate change [19, 20].

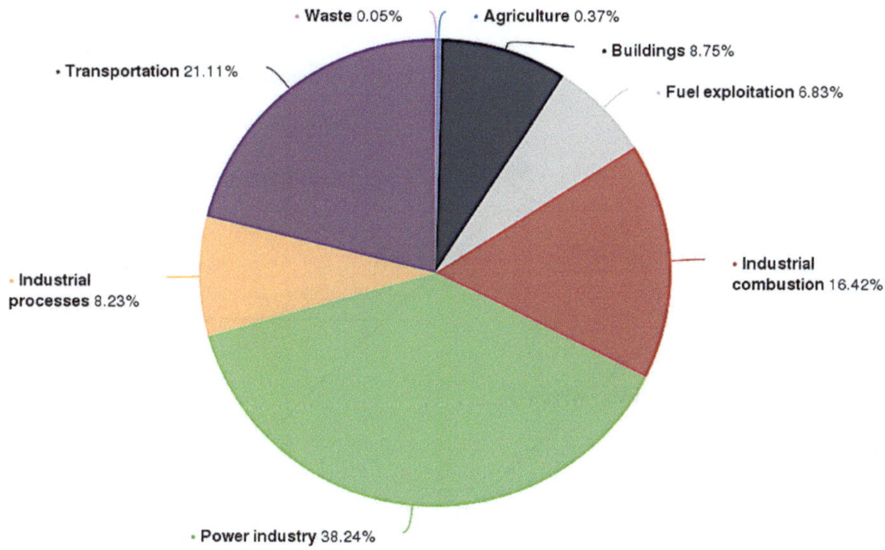

Fig. 1.4 Distribution of CO_2 emission worldwide in 2023, by sector. European Commission; EDGAR/JRC; Crippa et al. (2024); IEA; Statista, edgar.jrc.ec.europa.eu

1.3 Overview of CO_2 Sequestration

CO_2 sequestration, also known as carbon capture and storage (CCS), is a technique which can reduce the amount of carbon dioxide (CO_2) from the atmosphere and CO_2 released from industrial and energy-related sources. This aims to reduce the impact of greenhouse gases which create global warming and results in climate change [18–20]. This process has three main stages i.e. capture, transport, and storage (Fig. 1.5).

Capturing CO_2 emissions from industrial processes or power plants is the first stage in CO_2 sequestration [21, 22]. Numerous CO_2 capture technologies are used to separate CO_2 from other gases which are emitting during combustion and industrial works. These capture technologies can be categorised in to three main categories:

(i) Post-combustion CO_2 Capture: CO_2 emitted with the flue gases during the industrial process of fossil fuel emission, combustion processes in cement industry etc. CO_2 can be emitted from the flue gases after the combustion of fossil fuels and this process is known as post-combustion method. In this process, the flue gas is treated with the solvent that absorbs CO_2, which is then separated and captured. This approach is widely used in the power plants (pulverized coal (PC) plants and natural gas combined cycle (NGCC)) [19, 20].

(ii) Pre-combustion CO_2 Capture: In this process, the primary fuel is reacted with steam and air or oxygen to produce a mixture of carbon monoxide and hydrogen (syngas). The carbon monoxide is then converted to CO_2 in a shift reactor, after which the CO_2 is separated and captured. The remaining hydrogen can

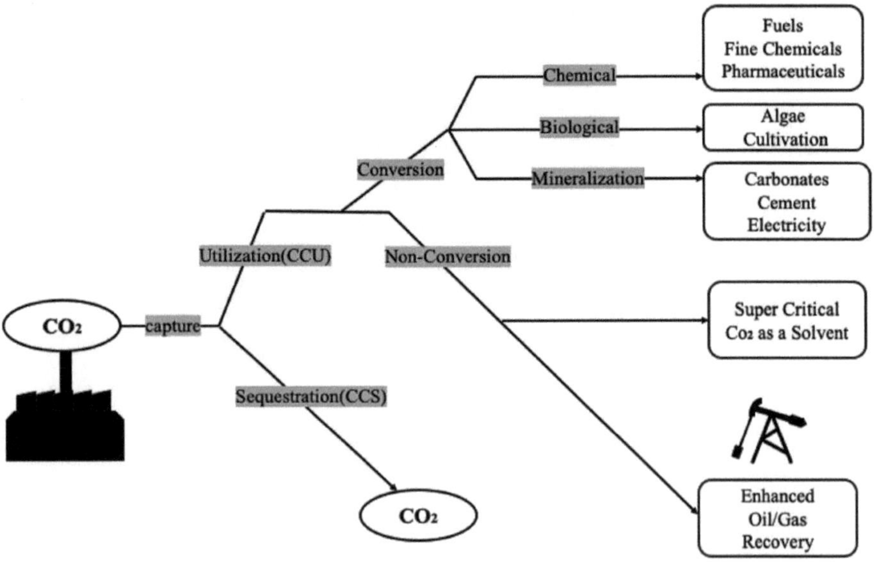

Fig. 1.5 Flow chart showing carbon capture and storage (CCS) process where a relatively pure stream of carbon dioxide (CO_2) from industrial sources is separated, treated and transported to a long-term storage location

be used to generate power. This method is particularly suitable for integrated gasification combined cycle (IGCC) power plants [20].

(iii) Oxy-fuel Combustion: As the flue gas consists primarily of both water and carbon dioxide, it can be feasible to differentiate carbon dioxide from the air by condensing the water. The ability to obtain up to one hundred percent of the carbon dioxide, or CO_2, and the fact that there are no emissions to the environment (no nitrogen oxides) are two of the main advantages of the process [21].

Once CO_2 has been captured, it must be shipped to a storage facility. Carbon dioxide is usually compressed to a fluid with a supercritical state, exhibiting features common to gases as well as liquids. This stage occurs with pressures across 7.4 MPa and temperatures exceeding 31 °C. Supercritical CO_2 is transported using pipelines, similar to those exploited for natural gas, or by ships if the storage location is isolated by water body. Pipelines for CO_2 transport are now functioning, particularly in the United States, wherein CO_2 is used for improved recovery of oil (EOR) [22]. CO_2 sequestration involves placing a quantity of carbon dioxide in an underground repository that ensures its secure and stable storage, limiting its escape into the atmosphere. This procedure is essential for reducing the carbon footprint and mitigating the effects of climate change. The most common alternatives for CO_2 sequestration are:

(i) Bio-sequestration describes the gathering and storing of CO_2 by living organisms, particularly plants and microorganisms, which absorb carbon via processes that include photosynthesis and store it in biomass or soils. This process, alluded to as biosequestration, consists the capture and storage of CO_2 by natural biological processes. Flora, comprising trees and other vegetation, capture CO_2 during photosynthesis and then store it inside of their biomass. Soil and wetlands provide as significant carbon sinks. Biologic sequestration can be enhanced through measures such as afforestation, reforestation, and improved land management that elevate the carbon storage capacity of soil and vegetation. This process involves various carbon sinks, notably forests, grasslands, soil, oceans, and other aquatic settings. It is regarded as a "indirect" or passive form of sequestration, as it relies on biological mechanisms to absorb and store carbon.

(ii) Oceans: atmospheric carbon dioxide can be solubilised in deep sea waters, where their re-emission into the atmosphere is less expected. This method is difficult owing to ocean acidification and its effects on marine ecosystems. Aquatic ecosystems, such as oceans and large water bodies, are significant absorbers of CO_2. The oceans absorb approximately twenty-five percent of the carbon dioxide emitted into the Earth's atmosphere. The greater part of this carbon is stored in the upper strata of the oceans. Excessive CO_2 absorption may result to water acidification, jeopardising marine biodiversity. This underlines the importance of minimising atmospheric CO_2 levels to protect the marine ecosystems.

(iii) Terrestrial Ecosystems: Carbon sequestration in terrestrial ecosystems involves capturing CO_2 through biological processes, such as tree planting (afforestation) or soil carbon storage in agricultural lands. Forests and woodlands are among the most effective natural carbon sinks. During photosynthesis, plants absorb CO_2 from the atmosphere and release oxygen, thus purifying the air. On average, forests store twice as much carbon as they emit. An estimated 25% of global CO_2 emissions are sequestered in forests and other vegetative forms, including grasslands and rangelands (fields, prairies, shrublands, etc.). Protecting these natural environments is crucial for maintaining their ability to capture CO_2 effectively. Deforestation and land-use changes, such as construction and intensive agriculture, pose significant threats to this natural process by reducing the number of plants available to absorb CO_2.

Soil acts as another important carbon sink, particularly in bogs, peatlands, and swamps. In these environments, CO_2 is captured and stored as carbonates, which build up over thousands of years as CO_2 combines with mineral elements like calcium or magnesium. While eventually, CO_2 can be released from the soil, this process takes a very long time—often more than 70,000 years in some cases.

(a) Underground Geological media: Geologic CO_2 Sequestration involves storing CO_2 in the geological rocks at different depth level. Depleted oil and gas fields, deep saline formations, and unmineable coal seams are considered as the suitable sites for storage. The CO_2 is injected to the subsurface geological formations and trapped by the processes like physical and chemical processes. Storing it in the subsurface media prevents it from escaping back into the atmosphere if the site is completely sealed naturally or artificially. Geological trapping mechanisms (e.g. structural, stratigraphic, residual, hydrodynamic, etc.) support the CO_2 geosequestration process.

1.4 Geological CO_2 Sequestration Options

Carbon Capture and Storage (CCS) is a complicated process where the acquires CO_2 is compressed and transferred to the storage sites with deep underground storage capacities, and injected into geological formations for permanent storage [11, 22]. This method is feasible to prevents CO_2 from entering the atmosphere, and to mitigate the global warming impacts. Effective CO_2 storage sites include depleted oil and gas reservoirs and deep saline formations, with an minimum depth of 800 m, where the temperature and pressures are sufficiently high enough to store the CO_2 in a liquid or supercritical state. These techniques prevents CO_2 from migrating from the storage sites through the physical, chemical and geophysical trapping mechanisms. The technologies used in oil and gas industries can be utilized for the CO_2 injection in the geological sites (Fig. 1.6). For understanding the storage capacity and monitoring the well-drilling and injection equipment, measurement and monitoring technologies are essential to observe the remaining capacity of the storage site and the behaviour of the CO_2. Although CO_2 injection technologies needs improvements. First CO_2 is injected into the reservoir and the well is sealed using cement 'plug,' at appropriate depth to restrict CO_2 from migrating up the well or escaping for contaminating groundwater.

CO_2 is injected into deep geological formations such as depleted oil and gas reservoirs, saline aquifers, or unmineable coal seams. These formations must have characteristics that prevent the CO_2 from escaping back to the atmosphere, such as impermeable cap rocks that trap the CO_2 underground. Once injected, the CO_2 is stored securely underground. Over time, it mineralizes or dissolves in underground fluids, becoming permanently trapped in geological formations [8, 9, 11]. Geosequestration is considered a costly process due to the energy required for capture, compression, and transport, as well as the infrastructure needed for injection and monitoring. However, it is seen as a crucial technology for reducing CO_2 emissions and mitigating climate change, especially in industries where direct emissions reductions are challenging to achieve. Biologic sequestration, on the other hand, involves using natural processes to capture and store carbon, such as through afforestation (planting trees) or restoring wetlands, which absorb CO_2 from the atmosphere and store it in biomass or soil organic matter. The final stage in CCS is the long-term storage of CO_2 in suitable geological formations.

Declined or abandoned oil and gas reservoirs are the storage sites that can be repurposed to store CO_2. As these fields already exists infrastructure and geological knowledge, they can be viable options for CO_2 storage [8]. Deep saline reservoirs are porous formations filled with saline water, and generally located at high depths. The saline reservoir at greater than 800 m are favourable for CO_2 storage. The high

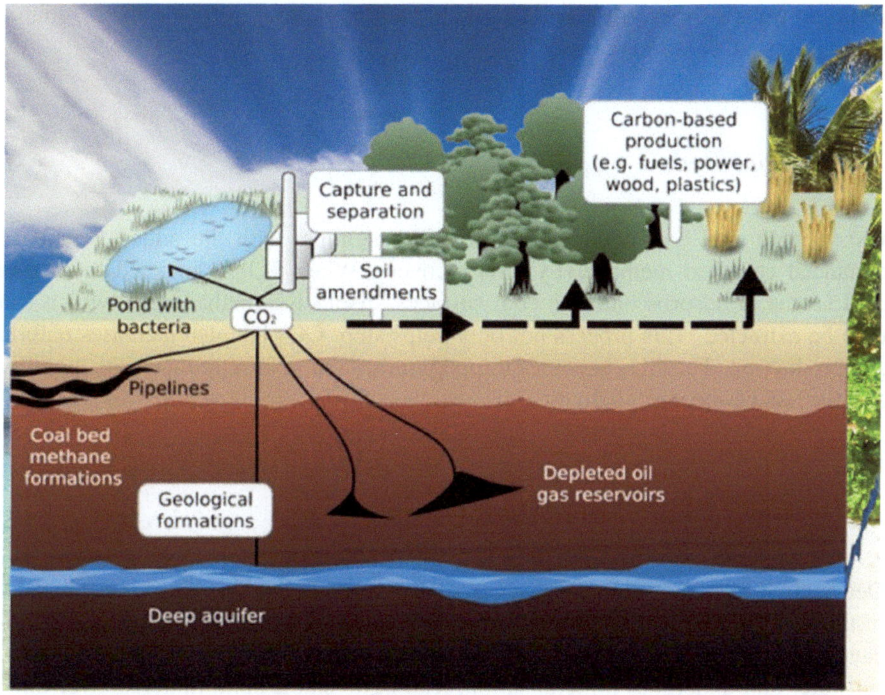

Fig. 1.6 Geological and biological sequestration

pressures and temperatures at high depths make the CO_2 a supercritical state, and ensured that it remains trapped. Coal seams are also the viable option for CO_2 storage, specially the unmineable coals [8, 9, 11]. CO_2 gets adsorbed onto the coal surface, and it displaces methane that can be captured and utilized. The storage storage capacity and entrapment systems are controlled by several factors and trapping mechanism to prevent CO_2 from seeping to the surface. These mechanisms include structural trapping by impermeable cap rocks, residual trapping within the pore spaces of the rock, solubility trapping where CO_2 dissolves in the brine, and mineral trapping where CO_2 reacts with minerals to form stable carbonates [8, 9, 22].

1.4.1 Depleted Oil and Gas Reservoirs

Depleted oil and gas reservoirs are one of the most promising storage site for CO_2. These reservoirs are already proven for trapping hydrocarbon gases for millions of years. Injecting CO_2 into these reservoirs may also enhance hydrocarbon recovery through Enhanced Oil Recovery (EOR) processes. In this process, the injected CO_2 helps in displacing the remaining oil, and improving extraction efficiency. The geologic traps of these reservoirs provide natural capping to make them reliable for long-term CO_2 storage [8, 20].

1.4.2 Coal Beds

Unmineable coal beds can be used for CO_2 storage, with the added benefit of enhanced coal bed methane (ECBM) recovery. When CO_2 is injected into these coal seams, it adsorbs onto the coal surfaces, displacing methane gas, which can then be extracted. This process not only helps store CO_2 but also increases natural gas production. The CO_2 becomes trapped in the coal matrix, reducing its mobility and the risk of seepage [11].

1.4.3 Deep Saline Formations

Deep saline formations are vast porous rock layers filled with saline water. In spite of their substantial size and wide availability, these types of structures have significant CO_2 storage potential. CO_2 injected through these structures dissolves into the saltwater and undergoes mineralisation, leading to stable solid minerals over time. Since saline deposits typically occur underneath the surface, they provide the necessary pressure and geologic barriers for maintaining CO_2 securely stored for many thousands of years [23].

1.4.4 Basalt

Basalt formations, which are made up of volcanic rock, offer another fascinating alternate for storing carbon dioxide because of their exceptionally high mineral reaction. Basalt's quick mineralisation makes it an appealing alternative for over time carbon storage, especially in areas with high volcanic rock formations [20]. Basalt sequestration entails the injection of CO_2 into the basalt traps and deep sea formations, where it at first interacts with seawater prior to reacting with basalt, a rock rich in alkaline elements. This reaction creates Ca_2^+ and Mg_2^+ ions, which lead to the formation of stable carbonate minerals. Basalt sequestration offers significant possibilities related to the broad availability of basalt and its ability for secure, long-term storage of carbon. This method uses natural processes for stabilising CO_2, providing a potential solution for enormous scale carbon storage [22–27].

1.4.5 *Mineral Sequestration*

A technique known as mineral sequestration entails the capture and storage of carbon in the manner of solid carbonate compounds. This natural process occurs over geological time, enabling the deposition and accumulation of limestone. Clay minerals are left as a residue after carbonic acid in groundwater performs a steady reaction with complex silicates. This process dissolves calcium, magnesium, alkalis, and silica amid other elements. Dissolved calcium and magnesium react with bicarbonate, leading to a precipitation of calcium and magnesium carbonates, a process utilised by organisms for creating shells. After the death of these organisms, their outer shells accumulate as sediment, ultimately transforming into limestone. Limestone accumulates over billions of years and contains an important proportion of Earth's carbon. Current research efforts focus on accelerating identical reactions in alkali carbonates to enhance mineral sequestration [25, 26].

1.4.6 *Advanced Materials for Carbon Capture*

Porosity, chemical stability, and resistance to heat are some of the features that are linked with zeolitic imidazolate frameworks (ZIFs), which are metal–organic frameworks that are equivalent to zeolites. ZIFs are being investigated for their capacity to capture CO_2 due to these properties. Carbon dioxide reacts exothermically with metal oxides, leading to the development of stable carbonates, such as calcite and magnesite. This natural process, leading substantially to the Earth's surface limestone, usually happens over a long time. Applying the properties of advanced materials such as ZIFs might enhance the efficiency and effectiveness of CO_2 capture and storage [28]. Continuous evaluation and confirmation is necessary to ensure the long-term safety and efficacy of CO_2 storage. Methods such as seismic surveys, pressure surveillance, and geochemical examination have been used to monitor the movement and activity of stored CO_2. This monitoring promotes the identification of potential leaks and the evaluation of the storing site's capacity and integrity.

1.5 Current Status

Several full-scale carbon capture and storage (CCS) projects are currently functioning globally, mainly within the industrial sector. The Weyburn project illustrates the utilisation of CO_2 obtained from a coal gasification facility in the United States. Approximately one million tonnes of carbon dioxide (CO_2) are being pumped to a deep saline formation underneath the North Sea by the oil corporation Statoil in Norway from the year 1996. CO_2 is obtained from natural gas. Since 2008, the Snohvit project in Norway has utilised similar technology. A consortium encompassing BP, Statoil, and Sonatrach has been involved with CO_2 the injection at the Salah site in Algeria, where the CO_2 is obtained from gas production. Technologies equivalent to capture before combustion are used in the production of fertilisers and hydrogen, resulting in the captured CO_2 being either used in many different processes or distributed into the surrounding environment. Oxy-fuel combustion technique for power generation stays in the demonstration stage and is currently in testing in Germany by Vattenfall, a European electricity supplier. [22–27].

Recent estimates show that the global capacity utilisation for geological CO_2 storage varies between 8000 and 15,000 gigatonnes (Gt) of CO_2. The comprehension of storage potentials varies across global, regional, and local contexts. Estimates of storage capacity are highest in Europe, North America, Japan, and Australia. Depleted oil and gas reservoirs hold an estimated global storage capacity of 675 to 900 Gt of CO_2, causing them an appealing choice due to the already present knowledge of these sites and the possibility of repurposing infrastructure from oil and gas extraction activities. Deep saltwater formations are thought to have a storage capacity of no less than 1000 Gt of CO_2 and are thoroughly distributed across numerous sedimentary basins globally. Additional information on storage capacities is essential in regions with rising electricity consumption, such as China, India, Southeast Asia, Eastern Europe, and Southern Africa. CO_2 sequestration has been carried out in many worldwide initiatives. The Sleipner gas field in Norway and the In Salah gas fields in Algeria illustrate successful CO_2 capture and storage efforts. The Quest CCS Project in Canada captures CO_2 from an oil sands upgrader and accumulates it in a deep saline formation. The Kemper County IGCC Project in the United States seeks to capture CO_2 from a power plant for enhanced oil recovery reasons. The projects described in Table 1.1 indicate the growing effectiveness and importance of technologies that capture carbon in reducing global CO_2 emissions.

The Global CCS Institute claims 12 functioning CCS projects as well as 8 additional projects presently under construction globally [29–31]. Major advances in CO_2 injection have been documented in formations such as the Bakken, Eagle Ford, Marcellus, and Barnett shales. These sites showed substantial enhancements in recovery rates, emphasising the potential for raised oil and gas recovery through CO_2 injection. The case studies show the dual benefit of decreasing the amount of CO_2 in the atmosphere while improving the efficiency of fossil fuel extraction.

1.6 Conclusion

Carbon Capture, Utilisation, and Storage (CCUS) contributes an essential part in achieving the world's climate targets. The significance of CCUS technologies goes beyond basic storage; that they promote the use of captured CO_2 in different industrial processes, which leads to a reduction in net emissions. CO_2 geosequestration is a crucial tool for mitigating climate change. Geological sequestration may make a significant contribution to meeting the ambitious targets established by the Paris Agreement by addressing vital factors for effective deployment and integrating with

Table 1.1 Example of carbon capture and storage projects worldwide [17, 23, 26, 29–31]

Project name	Year	Country	CO_2 storage estimation	Other details
Sleipner project	1996	Norway (North Sea)	>25 million tonnes of CO_2 stored	First commercial-scale carbon capture project. Demonstrated economic viability and large-scale potential
Gorgon project	Ongoing	Australia	Up to 4 million tonnes of CO_2 per year	Captures CO_2 from natural gas production. Faces criticism for high cost and delays but offsets emissions equivalent to over 800,000 cars annually
Petra nova project	2017	USA (Texas)	1.6 million tonnes of CO_2 per year	Captures 90% of CO_2 emissions from a power plant. Uses captured CO_2 for enhanced oil recovery (EOR). Supported by government grants and tax credits
Boundary dam project	2014	Canada	1 million tonnes of CO_2 per year	First commercial carbon capture on a coal-fired power plant. Stores 90% of emissions, equivalent to removing 250,000 cars
Quest project	2015	Canada	1 million tonnes of CO_2 per year	Captures emissions from bitumen upgrading. Known for low-cost and efficient design. Demonstrates potential for the oil and gas sector

other mitigation strategies. Achieving a limit on the worldwide temperature rise to 2 °C necessitate urgent and continuous action in carbon capture and storage (CCS) and other supporting technologies. Geological sequestration reflects an appropriate strategy for carbon storage; nevertheless, it makes up one component of an extensive range of strategies needed for solving climate change. The combination of geological sequestration alongside additional mitigation measures, such as the implementation of renewable energy and emissions reduction, is crucial for significant progress in decreasing the amount of CO_2 in the atmosphere.

References

1. Ritchie H, Roser M (2020) “CO_2 emissions”. Published online at OurWorldInData.org. Retrieved from: https://ourworldindata.org/co2-emissions [Online Resource]
2. Rummukainen M (2012) Changes in climate and weather extremes in the 21st century. Wiley Interdisc Rev Clim Change 3(2):115–129
3. Khan MZA (2012) Climate change: cause & effect. J Environ Earth Sci 2(4):48–53
4. Spiridonov V, Ćurić M, Spiridonov V, Ćurić M (2021) Climate and climate change. Fundam Meteorol 377–397

5. Meehl GA, Zwiers F, Evans J, Knutson T, Mearns L, Whetton P (2000) Trends in extreme weather and climate events: issues related to modeling extremes in projections of future climate change. Bull Am Meteor Soc 81(3):427–436
6. Lamb WF, Wiedmann T, Pongratz J, Andrew R, Crippa M, Olivier JG, Minx J et al (2021) A review of trends and drivers of greenhouse gas emissions by sector from 1990 to 2018. Environ Res Lett 16(7):073005
7. Von Blottnitz H, Curran MA (2007) A review of assessments conducted on bio-ethanol as a transportation fuel from a net energy, greenhouse gas, and environmental life cycle perspective. J Clean Prod 15(7):607–619
8. Singh S, Boruah A, Devaraju J (2023) CO_2 sequestration and CH_4 extraction from unmineable coal seams in Singrauli coalfield. Int J Coal Prep Utilization 1–18
9. Hazarika S, Boruah A, Saraf S (2023) Modeling and simulation study of CO_2 fracturing technique for shale gas productivity: a case study (India). Arab J Geosci 16(7):408
10. Statista. Distribution of Carbon Dioxide Emissions Worldwide in 2023, by Sector. Statista, Statista Inc., 5 Sep 2024, https://www.statista.com/statistics/1129656/global-share-of-co2-emissions-from-fossil-fuel-and-cement/
11. Hazarika S, Boruah A (2022) Supercritical CO2 (SCO_2) as alternative to water for shale reservoir fracturing. Mater Today: Proc 50:1754–1757
12. Salari M, Javid RJ, Noghanibehambari H (2021) The nexus between CO_2 emissions, energy consumption, and economic growth in the US. Econ Anal Policy 69:182–194
13. Menyah K, Wolde-Rufael Y (2010) CO_2 emissions, nuclear energy, renewable energy and economic growth in the US. Energy Policy 38(6):2911–2915
14. Onofrei M, Vatamanu AF, Cigu E (2022) The relationship between economic growth and CO_2 emissions in EU countries: a cointegration analysis. Front Environ Sci 10:934885
15. Lokhorst A, Wildenborg T (2005) Introduction on CO_2 Geological storage-classification of storage options. Oil Gas Sci Technol 60(3):513–515
16. Bachu S (2008) CO2 storage in geological media: role, means, status and barriers to deployment. Prog Energy Combust Sci 34(2):254–273
17. Zahid U, Lim Y, Jung J, Han C (2011) CO_2 geological storage: a review on present and future prospects. Korean J Chem Eng 28:674–685
18. Kelemen P, Benson SM, Pilorgé H, Psarras P, Wilcox J (2019) An overview of the status and challenges of CO_2 storage in minerals and geological formations. Front Clim 1:482595
19. Shahbazi A, Nasab BR (2016) Carbon capture and storage (CCS) and its impacts on climate change and global warming. J Pet Environ Biotechnol 7(9)
20. Anwar MN, Fayyaz A, Sohail NF, Khokhar MF, Baqar M, Khan WD, Nizami AS et al (2018) CO_2 capture and storage: a way forward for sustainable environment. J Environ Manag 226:131–144
21. Folger P (2013) Carbon capture and sequestration (CCS): a primer. Congressional Research Service, Washington, DC
22. Bui M, Adjiman CS, Bardow A, Anthony EJ, Boston A, Brown S, Mac Dowell N et al (2018) Carbon capture and storage (CCS): the way forward. Energy Environ Sci 11(5):1062–1176
23. DePaolo DJ, Cole DR (2013) Geochemistry of geologic carbon sequestration: an overview. Rev Mineral Geochem 77(1):1–14
24. Lackner KS (2002) Carbonate chemistry for sequestering fossil carbon. Annu Rev Energy Env 27(1):193–232
25. Saran RK, Arora V, Yadav S (2018) CO_2 sequestration by mineral carbonation: a review. Glob Nest J 20:497–503
26. Olajire AA (2013) A review of mineral carbonation technology in sequestration of CO_2. J Petrol Sci Eng 109:364–392
27. Sanna A, Uibu M, Caramanna G, Kuusik R, Maroto-Valer MM (2014) A review of mineral carbonation technologies to sequester CO_2. Chem Soc Rev 43(23):8049–8080
28. Ighalo JO, Rangabhashiyam S, Adeyanju CA, Ogunniyi S, Adeniyi AG, Igwegbe CA (2022) Zeolitic imidazolate frameworks (ZIFs) for aqueous phase adsorption–a review. J Ind Eng Chem 105:34–48

29. Huaman RNE, Jun TX (2014) Energy related CO_2 emissions and the progress on CCS projects: a review. Renew Sustain Energy Rev 31:368–385
30. Page B, Turan G, Zapantis A, Burrows J, Consoli C, Erikson J, Zhang T (2020) The global status of CCS 2020: vital to achieve net zero
31. Kearns D, Liu H, Consoli C (2021) Technology readiness and costs of CCS. Global CCS Institute 3

Chapter 2
Reservoir Screening Criteria for CO_2 Storage

Abstract This chapter gives an in-depth look at the ways that CO_2 is trapped and stored permanently in rock formations, as well as the factors that should be used to choose the best storage sites. The main trapping mechanisms consist of structural and stratigraphic trapping, which involves the physical confinement of CO_2 beneath impermeable caprocks; residual trapping, where CO_2 is immobilised in the pore spaces of the rock; solubility trapping, where CO_2 dissolves into formation fluids; and mineral trapping, where CO_2 interacts with minerals to form stable carbonates over time. While selecting geological formations for CO_2 storage, the most important variables for consideration are reservoir depth, porosity, permeability, and caprock integrity to ensure effective containment. This chapter discusses the importance of site-specific risk assessments, concentrating on potential leakage pathways and the long-term stability of storage sites. Different formation types, such as depleted oil and gas reservoirs, saline aquifers, and unmineable coal seams, are assessed for their suitability based on these criteria. This chapter emphasises that effective CO_2 storage is dependent on understanding the geological features of the site of storage and the mechanisms of trapping involved in it, all of which are essential to ensuring permanent confinement while minimising the adverse environmental effects.

Keywords CO_2 trapping · CO_2 storage · Structural trapping · Mineral trapping · Site selection

2.1 Introduction

Carbon dioxide (CO_2) capture and storage (CCS) is a key technology that can help mitigate the impacts of climate change by limiting the emission of CO_2 into the atmosphere [1, 2]. Geological storage, whereby captured carbon dioxide is injected into deep subsurface rocks, has become one of the most exciting options for the long-term

The original version of the chapter has been revised. A correction to this chapter can be found at https://doi.org/10.1007/978-3-031-81021-3_8

A. Boruah, *CO_2 Geosequestration: Capturing Carbon for a Sustainable Future*, Springer Climate, https://doi.org/10.1007/978-3-031-81021-3_2

sequestration of carbon dioxide. In this chapter, we delve into the mechanisms that are accountable for the underground storage of carbon dioxide (CO_2) in addition to the fundamental selection criteria that are employed in determining geological formations which are suitable for storage. Mechanisms that trap carbon dioxide are crucial processes that ensure the safe and comprehensive storage of carbon dioxide in geological formations over a long period of time [3–5]. Structure-based trapping is one of these mechanisms (Fig. 2.1), which is a process whereby carbon dioxide is physically trapped underneath an impermeable caprock, so restricting its upward movement. Solubility trapping involves the dissolution of carbon dioxide in formation water, resulting in it less buoyant and more stable. Residual trapping, on the contrary, entails immobilising carbon dioxide in the pore spaces of reservoir rocks using the capillary forces. This process of mineral trapping occurs over time when dissolved carbon dioxide incorporates with minerals in the rock that contains it, leading to the development of stable carbonates that permanently lock carbon dioxide in solid form [5–8]. A multi-layered method of safely storing carbon dioxide and limiting its release into the atmosphere is provided by these systems, which act together to generate this approach. Using three main trapping mechanisms, supercritical carbon dioxide that has been injected into subsurface formations is safely trapped. These techniques are as follows: (1) chemical trapping, (2) physicochemical trapping, and (3) physical trapping. In order to ensure that CO_2 is contained for an extended period of time, the total efficiency of CO_2 storage is dependent on the combined activity of these processes [5–9]. In order to comprehend the dynamics of these trapping approaches, it is necessary to have a matrix that depicts how they interact with one another. A number of similarities exist between chemical and physical trapping, particularly in terms of the physicochemical processes involved. Hydrodynamic trapping is a type of physicochemical trapping that takes place on both the chemical and physical levels [3–9].

Carbon capture and storage (CCS) programs intend to mitigate the effects of climate change, and one of the most important aspects of these initiatives is the screening of carbon dioxide (CO_2) storage in geological formations. For the purpose of determining whether or not potential geological reservoirs, such as depleted oil and gas fields or deep saline aquifers, are suitable for the long-term storage of carbon dioxide, this procedure entails the careful investigation and assessment of this potential geological reservoir. During the screening procedure, a number of geological, hydrogeological, and geophysical parameters are taken into consideration in order to guarantee the storage site's integrity and safety. For the purpose of determining the containment capability of the selected geological formation, it is important to take into consideration a number of variables such as rock permeability, porosity, and caprock integrity [5, 6]. In addition, the process of screening takes into account a number of important aspects, including proximity to pollution sources, transportation infrastructure, and legal considerations. A important component in the global effort to minimise greenhouse gas emissions and battle climate change is the selection of safe and efficient storage locations, which can be achieved through the careful evaluation of geological formations [5–9].

Fig. 2.1 CO_2 trapping mechanisms

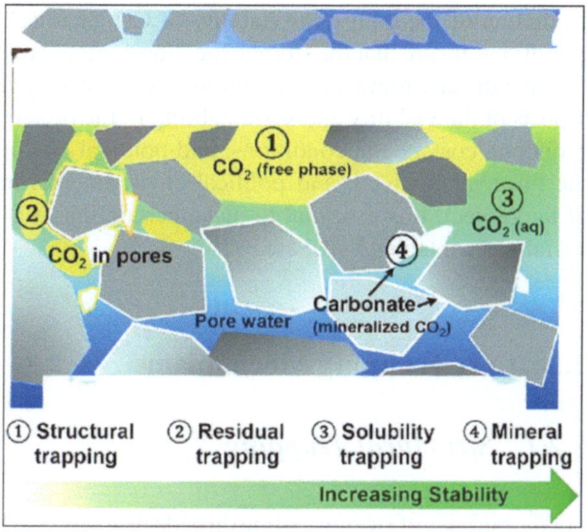

The selection criteria for CO_2 storage in geological formations include several essential screening points, such as the category of sedimentary basin (the formation and characteristics of reservoirs are influenced by the varying geological and tectonic settings of different types of sedimentary basins). It is essential to have a solid understanding of these distinctions in order to identify appropriate storage locations. (i) Characterisation of the Reservoir: The depth and overall dimensions of the reservoir are important elements to consider when determining the storage capacity of the reservoir and designing extraction procedure. It is generally true that reservoirs that are deeper and larger in size offer more storage potential. (ii) The thickness of the reservoir (the diameter of the reservoir has an impact on the volumetric potential of the reservoir for the storage of fluids). Larger amounts of carbon dioxide can be stored in reservoirs that are thicker. (iii) Fractures of the tectonic and geological systems (the presence of tectonic and geological fractures is necessary for determining the structural integrity of the reservoir and locating potential pathways for fluid flow that could result in leakage). The petrophysical, geochemical, and geomechanical properties of the reservoir rock (these properties provide insights into the composition, stability, and behaviour of the reservoir rock, which helps in the development of suitable storage ideas). (iv) Thermal Maturity (are connected via an impact on the behaviour and characteristics of the fluids that are stored, as well as the type and quality of the fluids that are stored). This element is crucial for obtaining an understanding of how carbon dioxide will react with the environment of the reservoir. (v) Flow behaviour (the flow of fluids inside the reservoir is crucial for improving extraction and injection approaches, which in turn makes sure that CO_2 storage remains efficient and safe). The hydrogeological and geothermal gradient, which takes into account the flow and behaviour of fluids within the reservoir, as well as the possibility for heat-related affects evaluation, is a factor that contributes to the prediction of the long-term viability of CO_2 that has been deposited. The weather conditions ensure that the storage location continue to be achievable regardless of the fact that the environmental conditions are constantly changing. It is necessary for all aspects of the viability and sustainability of the project to take into consideration the economic cost, social conditions, and political conditions. These variables involve the financial, societal, and political environments. In order to ensure the success of the project, it is essential to make certain that it is both financially possible and socially acceptable [5–10] etc.

2.2 CO_2 Trapping Mechanisms

2.2.1 *Structural Trapping*

For the purpose of storing carbon dioxide in geological formations, structural trapping is one of the most fundamental types of mechanisms. It takes place when carbon dioxide is introduced into porous rock formations before being physically trapped beneath an impermeable layer, which is frequently referred to as a caprock or seal all through this process [5–9]. Because it is less dense than the fluids that are surrounding it, carbon dioxide has an ability to travel higher within the formation. However, its movement is impeded by the caprock, which acts as a barrier and prohibits it from escaping to the surface. Structural trapping works because of a number of important features, include the following:

(a) Caprock Integrity: The caprock must be impermeable and free of fractures or flaws that could allow carbon dioxide to escape. Shale, mudstone, and clay are all examples of common caprocks, and all of these varieties have a very low permeability.
(b) Geological Structure: The most important aspect to consider is the geometry of the subsurface deposit. Geological features that include anticlines, which are structures that fold upwards, fault traps, and salt domes are examples of phenomena that serve as natural traps for carbon dioxide.
(c) At suitable depths (typically greater than 800 m), the pressure is high enough to hold carbon dioxide in a supercritical situation. This leads to the carbon dioxide to grow more dense and less mobile, which in turn increases the efficiency of the trapping process.
(d) Injection Rate and Volume: The rate at which carbon dioxide is injected and the total volume can have an impact on how the carbon dioxide plume spreads and interacts with the geological structure, which in turn can have an effect on how successfully it confines itself.

As an outcome of its buoyancy, carbon dioxide (CO_2) starts moving upwards when it is injected into a porous rock layer, such as sandstone or carbonate. This causes the formation fluids, such as brine, to be displaced. As it rises, carbon dioxide build-up occurs beneath the caprock, which results in the formation of a structural seal. Whether it be an anticline, faulted trap, or stratigraphic trap, the geometry of the geological structure is responsible for directing the migration of carbon dioxide, which in turn leads it to pool and become restricted beneath the impermeable layer [8, 9].

As result of the reality that CO_2 occupies a less volume in its supercritical state compared to its gaseous state, it is possible to store it more effectively. In order for the structural trapping mechanism to be effective, it is necessary for the caprock to be able to maintain its integrity and maintain its resistance to the pressure that is being applied by the CO_2. It is possible that additional trapping mechanisms, such as residual, solubility, and mineral trapping, will further immobilise the carbon dioxide over time; nonetheless, structural trapping is responsible for the initial containment [7–9].

One of the most significant instances of structural trapping in action is the Sleipner Project, that is located in the North Sea and is the world's first large-scale carbon dioxide storage project. A deep saline aquifer that lies beneath a thick caprock of shale is known as the Utsira Formation. Since 1996, more than 25 million tonnes of carbon dioxide have been injected into this formation. The Utsira Formation is a strong candidate for structural trapping due to its structure, which consists of a thick and continuous shale caprock, as well as its excellent porosity and permeability. Over the course of several decades of operation, the caprock has been able to successfully contain the carbon dioxide and prevent any leakage to the surface.

The study of structural trapping extends back several decades, particularly in the context of oil and gas exploration. It was in this context that the understanding of how hydrocarbons are trapped in reservoirs provided the framework for the storage of carbon dioxide [5] or [6]. Research conducted in the past on geological formations such as fault traps and anticlines has provided valuable insights into the behaviour of buoyant fluids when subjected to pressure in porous media. Research on carbon dioxide storage gained momentum in the 1990s with the implementation of pilot projects such as Sleipner and following studies that tried to characterise geological formations that could be used for carbon dioxide storage. Researchers that employed seismic imaging and modelling to track the migration of carbon dioxide and the long-term stability of storage sites emphasised the significance of maintaining the integrity of the caprock throughout the process. Experimental and field data from initiatives such as In Salah (Algeria) and Weyburn-Midale (Canada) contributed further to the knowledge of structural trapping mechanisms. These projects occurred in Algeria and Canada, respectively. As of today, structural trapping continues to be an essential mechanism for the storage of carbon dioxide in geological formations. The efficiency of this mechanism has been demonstrated by a number of active initiatives all around the world. Using structural traps, large-scale projects such as Sleipner, Gorgon (Australia), and Quest (Canada) continue to store millions of tonnes of carbon dioxide (CO_2). At the moment, the primary focus of research is on enhancing monitoring methods, such as four-dimensional seismic imaging, in order to guarantee that the carbon dioxide will continue to be safely contained. The development of computer modelling is assisting in the prediction of how injected carbon dioxide would behave over the course of years, which increases the level of understanding in the long-term stability of storage sites [5–11]. Despite this, there are still obstacles to overcome, notably with regard to the identification of suitable geological structures that contain intact caprock. Understanding the possibility for leaking along faults or fractures is an important topic of study since it is a serious concern that is posed by caprocks that have been fractured or that have failed. Additionally, induced seismicity from CO_2 injection has been seen in certain regions, which has raised concerns about the integrity of caprocks in regions that are geologically active [5–8].

2.2.2 Residual Trapping

CO_2 is trapped within the pore spaces of a geological formation by a process known as residual trapping, which is an important mechanism in the process of CO_2 storage. It is due to the capillary forces that exist inside the rock that residual trapping takes place, which prevents carbon dioxide from moving through the formation. When injected carbon dioxide is driven into the pore spaces of a rock formation, but the brine (formation water) that is already present inside the formation is prevented from being fully displaced, this phenomenon takes place. Capillary pressure causes little "bubbles" or ganglia of carbon dioxide to remain immobilised as it migrates through the porous medium [5–8, 12, 13]. This occurs because carbon dioxide becomes caught in the narrow pore spaces as it continue to move through the medium.

Additionally, the wetting characteristics of the rock have an effect on the effectiveness of the residue trapping process. It is common for the rock in saline aquifers to be water-wet, which suggests that brine has a tendency for occupying the smaller pore spaces, while carbon dioxide must go through the larger pores. This leads to the brine to be moved while the CO_2 is injected; however, once the injection ceases, the capillary forces motivate the brine through and back into the smaller pores, which then causes the CO_2 to stay in its original location. It is no longer feasible for this trapped carbon dioxide to move, which essentially locks it down in the subsurface for extended periods of time (5), (6), (7), (8), (11), and (12). There is a correlation between the relative permeability of carbon dioxide and water in the formation and residual trapping. Whenever there is a reduction in the quantity of free-flowing carbon dioxide, the relative permeability to carbon dioxide falls, which further enhances its immobilisation. Through the use of this method, the risk of CO_2 leakage is decreased, particularly in formations that do not possess structural traps.

The efficiency and degree of residual trapping are both controlled by a number of parameters, including the following:

(a) Pore Size and Distribution: The size of the pores within the rock as well as their distribution are factors that are extremely important for residual trapping. The smaller cracks in rocks with a wide variation of pore sizes are more likely to capture and hold carbon dioxide due to the capillary forces that are present. This allows the rocks to more efficiently trap carbon dioxide [2, 6] and [10].

(b) Wettability: The likelihood that a fluid may spread across or stick to a solid surface is what is meant by the term "wettability." The presence of brine in water-wet formations results in a greater affinity for the rock surface. This, in turn, increases the displacement of carbon dioxide into larger pores, which in turn enables more effective concentration. It is vital to have the wetting characteristics of the reservoir in order to achieve the highest possible residual trapping [8, 14, 15].

(c) Injection Rate: The rate at which carbon dioxide is supplied to the formation has an effect on the manner in which the gas travels through the pore spaces. When analysing the performance of storage, the standard benchmark for the rate of CO_2 injection is typically one million tonnes per year. As a result of CO_2 being able to flow through pores that are smaller, faster injection rates may result in less effective residual trapping. On the other hand, slower injection rates can lead to a more uniform dispersion of carbon dioxide and increased trapping efficiency.

(d) Capillary Pressure: The capillary pressure is an important component that plays a role in determining the degree to which carbon dioxide is immobilised. It is possible for a higher capillary pressure to improve the efficacy of the trapping process by driving carbon dioxide into smaller pore spaces. This makes it more difficult for the gas to migrate once the injection has been halted.

The concept of formation permeability refers to the degree to which fluids are able to pass through a rock formation with relative ease. When formations have a low permeability, the transport of carbon dioxide is slower, which gives it more time to become trapped by capillary forces. This is one reason why low-permeability formations tend to support larger levels of residual trapping. Alternatively, in formations with a high permeability, carbon dioxide may be able to move about more freely, hence minimising the amount of residual trapping.

A location in Canada known as the Weyburn-Midale Project, which was initially an enhanced oil recovery (EOR) process, has provided evidence that residual trapping can play a role in the long-term storage of carbon dioxide gases. The monitoring that takes place after the injection has shown that a significant amount of the carbon dioxide that is injected is able to become trapped in the pore spaces of the reservoir rock, which contributes to the overall stability of the energy storage location.

There has been a significant amount of research conducted on residual trapping, conducted in both field initiatives and laboratory tests. Initially, the focus of research was on gaining an understanding of how capillary forces influence the flow of carbon dioxide through porous media. Studies that used core samples and simulations of reservoirs demonstrated that residual trapping has the potential to immobilise anywhere from ten percent to thirty percent of the injected carbon dioxide, depending on the characteristics of the rock and the fluid dynamics.

Several important laboratory investigations have investigated the impact that wettability and pore size distribution have on the process of residual trapping. In example, tests conducted with various types of rocks, such as sandstone and carbonate, have provided insights into the potential for trapping that is present in a variety of geological formations. The results of these investigations have been used to inform the creation of improved CO_2 storage models, which now take into account residual trapping as a significant component in forecasting the consequences of long-term storage.

In recent years, there has been a major advancement in the knowledge of residual trapping. This advancement has been made possible by improved modelling tools and monitoring technology, which have enabled more precise predictions of the behaviour of carbon dioxide in the subsurface. These techniques have enabled researchers to measure the amount of carbon dioxide that remains trapped in formations over time. Studies conducted in the field, such as those conducted at Sleipner and Weyburn-Midale, continue to demonstrate the significance of residual trapping strategies. Additionally, continuing research is in the process of investigating approaches to reduce residual trapping by optimising injection strategies and improving the characterisation of reservoir features. In addition, developments in digital rock physics and pore-scale modelling are delivering fresh insights into the ways in which residual trapping might be maximised in a variety of geological formations. With the use of these technologies, scientists are able to model the behaviour of carbon dioxide (CO_2) at a microscopic level, which provides a better knowledge of how capillary forces behave in storage scenarios that occur in the real world.

2.2.3 Solubility Trapping

A phenomenon known as solubility trapping makes it possible for injected carbon dioxide to dissolve into the brine or water of the formation, hence lowering the formation's buoyancy and mobility. In the process of geological storage, carbon dioxide is introduced into deep saline aquifers, where the high pressure and temperature cause it to reside in a condition that is referred to as supercritical. When carbon dioxide (CO_2) comes into contact with formation water, it begins to dissolve over time, resulting in the creation of carbonic acid (H_2CO_2). It is less likely that this dissolved carbon dioxide will move towards the surface since it no longer exhibits the characteristics of a buoyant gas. Over the course of time, the brine that is rich in carbon dioxide gets heavier than the formation water that is surrounding it. This causes the brine to sink deeper into the reservoir, which is a process known as convective mixing. This process improves the long-term stability of the carbon dioxide [16, 17]. The process of solubility trapping is a slow one, often taking place over a period of years to decades, but it makes a major contribution to the overall security of the storage location. Due to the fact that dissolved carbon dioxide is less likely to escape than free-phase carbon dioxide that is trapped under a caprock, this method is considered to be one of the most stable and long-lasting trapping mechanisms [16, 17] Is.

The efficacy of solubility trapping in geological formations is influenced by a number of factors, including the following [16–20].

(a) Temperature and Pressure: The solubility of carbon dioxide in brine decreases with temperature and increases with pressure. A greater amount of carbon dioxide can dissolve in brine when it is present in deep aquifers, which are characterised by high pressure and relatively low temperatures. Because of this, deeper storage formations often offer a greater potential for the entrapment of solubility concentrations.
(b) The Salinity of the Formation Water: The coefficient of CO_2 solubility has a negative correlation with the salinity of the formation water. The higher the salinity of the water, the lower the CO_2 concentration that can be dissolved. A large amount of carbon dioxide (CO_2) can still dissolve over time, especially in formations that are extremely salty, which is a factor that contributes to the safety of long-term storage.
(c) Composition of the Brine: The chemical compounds that make up the brine are an essential component in the process of dissolving carbon dioxide. Additionally, the presence of ions such as calcium, magnesium, and bicarbonates can have an effect on the rate of dissolution and the subsequent mineral entrapment that occurs.
(d) Properties of the Reservoir Rock: The porosity and permeability of the reservoir rock have an effect on the flow and mixing of carbon dioxide and brine. Formations with a high permeability allow for improved contact between carbon dioxide and brine, which in turn promotes effective dissolution. Furthermore, formations that have high vertical and lateral continuity restrict carbon dioxide from escaping, which allows for more time for dissolution to take place.
(e) Convection Driven by Density: When carbon dioxide is dissolved in brine, the amount of carbon dioxide in the water increases, which results in the formation of convection currents that move downward. The mixing is improved as a result, which further encourages the breakdown of free-phase carbon dioxide into the brine. Reservoir heterogeneity, which creates natural convective flow patterns, has the potential to make this process more effective.

Sleipner Project, which has been running in the North Sea since 1996, is a renowned example of solubility trapping in action. This project has been in operation since 1996. During the process of injecting carbon dioxide into a salty aquifer, a sizeable amount of the carbon dioxide has dissolved into the formation water over the course of time. Monitoring activities, such as four-dimensional seismic surveys, have revealed that the carbon dioxide plume has been steadily spreading and interacting with the brine. This suggests that solubility trapping is playing a role in the long-term containment of carbon dioxide in the reservoir. The Sleipner Project is a real-world example of how solubility trapping can be used in concert with other trapping processes to get the desired results.

To gain a better understanding of the function that solubility trapping plays in carbon sequestration, researchers have conducted substantial research on the topic. In the 1990s, the first models of CO_2 dissolution processes were produced. These models concentrated on fundamental dissolution kinetics as well as parameters such as brine composition and pressure conditions. At the beginning of the twenty-first century, researchers started looking into convective mixing and the ways in which the density differences between CO_2-rich brine and CO_2-free brine cause the downward movement of CO_2-enriched water [19]. Researchers Ennis-King and Paterson (2003) conducted a series of noteworthy investigations in which they modelled convective mixing and shown that solubility trapping might make a major contribution to the stability of long-term CO_2 storage. Since that time, research has expanded, and more sophisticated simulation methods have been utilised in order to get a deeper comprehension of the effects that geological heterogeneity, brine composition, and thermodynamic features have on solubility trapping [18, 19]. Solubility trapping is widely acknowledged as a crucial mechanism in CO_2 storage at the present time. It is one of the mechanisms that contributes to the overall security of storage sites, along with structural trapping, residual trapping, and mineral trapping processes. Research that is now being conducted is centred on enhancing the predictability of solubility trapping by means of improved reservoir modelling and field data from operational carbon capture and storage projects [16–18]. There has been an increase in the accuracy of tracking CO_2 transport and dissolution within storage facilities as a result of technological improvements such as fluid sampling and four-dimensional seismic monitoring four-dimensional. As an illustration, the continuing research conducted at Sleipner and Snøhvit in Norway has demonstrated that the process of solubility trapping plays a significant role throughout the duration of these initiatives. In addition, more advanced numerical models currently incorporate a variety of parameters, including as chemical reactions, brine chemistry, and geomechanical behaviour, in order to make more precise predictions regarding CO_2 dissolving rates. The operators are able to optimise storage tactics with the assistance of these models, which maximise solubility trapping and long-term CO_2 containment.

2.2.4 Mineral Trapping

The sequestration of carbon dioxide (CO_2) in geological formations typically involves mineral trapping, which is considered to be one of the most reliable and long-lasting methods. As a result of this process, carbon dioxide (CO_2) is immobilised for geologic time periods by the formation of stable carbonate minerals, which are formed when CO_2 combines with minerals in the host rock. There is a permanent solution to the problem of CO_2 storage that can be achieved by this method of trapping. This is because the CO_2 is incorporated into the mineral structure, making it difficult for it to be released back into the atmosphere [19, 20]. A number of factors, including the following, have an impact on the efficiency and rate of mineral trapping:

(a) Composition of the Host Rock: The mineral composition of the host rock is the most important factor to consider. Mineral trapping is best accomplished with rocks that are abundant in calcium, magnesium, and iron silicates. Examples of such rocks include basalt, peridotite, and ultramafic rocks. This is because these elements react with carbon dioxide to generate stable carbonates. Although sandstones and carbonates are porous, it is possible that they are not as reactive as rocks that are rich in silicate [19].

(b) Temperature and Pressure: The circumstances of temperature and pressure which exist within the geological formation play a vital effect in the kinetics of mineral entrapment. A higher temperature, which is often found in deeper formations, speeds up the chemical reactions that take place between carbon dioxide and the minerals that make up rock. As a result of its ability to diffuse more easily into the rock matrix, supercritical carbon dioxide, which is present at high pressure and temperature, also contributes to the enhancement of these processes [19, 20].

(c) CO_2 Concentration: The amount of carbon dioxide that is present in the pore fluid has an effect on the rate at which minerals are trapped. As a result of the increased interactions with reactive minerals that are facilitated by higher concentrations, the creation of carbonates is accelerating. On the other hand, mineral trapping may take place at a slower rate in formations that have lower concentrations of carbon dioxide [21].

(d) Fluid Chemistry: The presence of formation water, also known as brine, has a substantial impact on the process of mineral trapping. When silicate minerals are dissolved in saline solutions, they are more likely to react with carbon dioxide, which results in the formation of carbonates. A further factor is the pH of the formation water, which plays a role in the process. Conditions that are more acidic facilitate the dissolution of host minerals, which in turn makes their ions accessible for reaction with carbon dioxide.

(e) Surface Area of the Reaction: The surface area of the reactive minerals that are exposed to CO_2 as well has an effect on the efficiency of the trapping process. In general, rocks that have a higher porosity and permeability, as well as rocks that have been fractured, have a greater surface area that is available for reactions, which can lead to an increase in the rate at which carbonate compounds are formed [10, 11, 21].

Mineral trapping is accomplished through a series of chemical reactions that take place between carbon dioxide and the minerals that are present in the host rock. They can be broken down into three primary stages, which are as follows: CO_2 is dissolved into the formation water as the initial step in the process [19–21]. Carbon dioxide (CO_2) is dissolved in brine, resulting in the formation of carbonic acid (H_2CO_2), which decomposes into bicarbonate (HCO_2^-) and hydrogen ions (H_2O). As a result, an acidic environment is produced, which serves to facilitate the disintegration of silicate minerals. Silicate minerals including olivine, pyroxene, and feldspar are dissolved in the second step of the process, which is carried out by the acidic formation water. It is during the process of dissolution that calcium (Ca_2^+), magnesium (Mg_2^+), and iron (Fe_2^+) ions are released into the water. Numerous elements, including temperature, pressure, and the presence of reactive minerals, all have a role in determining the pace at which minerals dissolve. The precipitation of carbonate minerals is the fourth and last step in the process. Calcite ($CaCO_2$), magnesite ($MgCO_2$), and siderite ($FeCO_2$) are examples of stable carbonate minerals that are formed when the calcium, magnesium, and iron ions that are released during mineral dissolution combine with the bicarbonate ions (HCO_{23}^-) that are present in the brine. These carbonate minerals are of a solid and stable nature, and they effectively lock the carbon dioxide into an equilibrium that is permanent [19–21].

The overall process can be represented by the following simplified reaction:

$$CO_2 + H_2O \rightarrow H_2CO_3 \rightarrow H^+ + HCO_3^-$$
$$Ca^{2+} + HCO_3^- \rightarrow CaCO_3(\text{calcite})$$

This sequence of dissolution and precipitation reactions can take thousands of years, but it results in the most permanent form of CO_2 sequestration.

One of the most well-known examples of mineral trapping that has been successful is the CarbFix project in Iceland. CO_2 that has been dissolved in water is injected into basaltic rocks as part of this experiment. The CO_2 then combines with the calcium and magnesium that are present in the basalt to produce carbonates. CarbFix has showed that up to 95% of the injected CO_2 may be mineralised within two years, which is an exceptional pace when compared to other types of geological formations. This is due to the high reactivity of basalt, which speeds up the mineralisation process. The success of CarbFix demonstrates the potential for mineral trapping in basalt formations, particularly in areas where reactive rocks that are ideal for the purpose are present. As a result of this effort, future mineral trapping endeavours in geological settings that are comparable to this one have been made possible.

Extensive research has been carried out on the mechanisms that are responsible for mineral trapping in a variety of geological environments. Experiments conducted in the laboratory, research conducted in the field, and computer simulations have all contributed to a better understanding of the conditions during which mineral trapping takes place. The kinetics of CO_2–mineral reactions were the primary focus of the initial laboratory investigations on mineral entrapment. These experiments were conducted under controlled settings. The results of these investigations indicated that the process of mineralisation may be accelerated by elevating the temperature, pressure, and acidity of the fluid. By conducting experiments with core samples taken from basaltic and peridotite formations, researchers were able to gain a better understanding of the rates at which silicate minerals dissolve and the development of carbonates throughout the course of time. Important information regarding the ways in which carbon dioxide interacts with various types of host rocks was provided by the In Salah Project in Algeria. The monitoring of the long-term behaviour of carbon dioxide in the subsurface, particularly its interaction with rock minerals, was the primary focus of these large-scale CO_2 injection projects. Despite the fact that these initiatives were primarily concerned with structural and residual trapping, they made significant contributions to the development of future mineral trapping research [21]. The simulation of CO_2-mineral interactions in geological formations has been accomplished through the use of numerical models. These models assist in predicting the long-term fate of CO_2 that has been injected, including the amount of CO_2 that will be frozen in the form of minerals. Especially in formations that are abundant in reactive minerals, mineral trapping has been shown to be capable of accounting for a sizeable fraction of carbon dioxide (CO_2) sequestration, as demonstrated by simulations [21].

The use of mineral trapping as a dependable and long-term option for carbon dioxide (CO_2) sequestration is gaining popularity, particularly in areas that have geological formations that are ideal for the purpose. Optimisation of mineral trapping techniques and expansion of their application are the goals of a number of active research projects and projects that are now underway. Several other initiatives all over the world are investigating the possibility of mineral trapping, in addition to the CarbFix project that is being carried out in Iceland. In Washington, United States of America, the Wallula Basalt Pilot Project is conducting research on the mineralisation of carbon dioxide (CO_2) in basalt formations. Meanwhile, other studies are concentrating on the potential of ultramafic rocks for rapid mineral entrapment. There are currently ongoing research endeavours aimed at gaining a deeper comprehension of the reaction kinetics involved in mineral entrapment, particularly in a variety of geological environments. Additionally, there is a rising interest in boosting the mineralisation process by introducing catalysts or optimising fluid chemistry in order to speed up CO_2-mineral interactions.

The natural rate of mineralisation in most formations is relatively sluggish, which is one of the primary obstacles that must be overcome in order to scale up mineral trapping. On the other hand, recent developments in our knowledge of rock chemistry, fluid dynamics, and temperature control point to the possibility of speeding up this process. In addition, the availability of suitable rock types is restricted regionally, which may make it difficult to implement large-scale deployment in specific places.

2.2.5 Trapping via Hydrodynamic Forces

The manipulation of carbon dioxide (CO_2) in fluid systems by the application of hydrodynamic forces is accomplished through the process of hydrodynamic CO_2 trapping. When it comes to the efficiency of CO_2 trapping, the flow rate and velocity of the carrier fluid are important factors. Increasing the velocity of the CO_2 can improve its mixing and transport, but it may also have an effect on the stability of the trapping. Whether a flow pattern is laminar or turbulent can have an effect on how carbon dioxide is dispersed and trapped. In microfluidics, laminar flow is frequently utilised for the purpose of regulated trapping. The degree to which CO_2 interacts with the flow of the fluid and the effectiveness with which it is trapped is influenced by the viscosity of the carrier fluid. CO_2 retention can be improved by using fluids with a higher viscosity. Both buoyancy and separation efficiency are impacted by the density of the fluid in comparison to the CO_2 concentration. The amount of carbon dioxide that can be dissolved and then trapped is directly proportional to the solubility of carbon dioxide in the carrier fluid. It is possible that a higher solubility will result in a more effective trapping [19–21].

2.3 Key Considerations for CO_2 Storage Site Selection

There are a number of important factors to take into account when selecting appropriate locations for the storage of carbon dioxide (CO) in order to guarantee that the process is dependable, efficient, and safe for the environment. Key considerations to take into account while identifying suitable CO_2 sequestration locations in geological formations include the following:

2.3.1 The Capacity or Storage

"Storage potential" is a word that describes the total volume that is contained within a geological formation and has the potential to be utilised for the storage of chemicals such as carbon dioxide. The usual approaches for evaluating this capacity in the context of gas reservoirs frequently do not take into account the function that formation water plays, instead concentrating solely on the volume that is available after or during the process of gas extraction [22, 23]. It is possible that this technique is unduly conservative because it may fail to take into account the possibility that carbon dioxide may dissolve in the formation water. This is a component that has the ability to considerably boost storage potential. Both the subsurface pressure and temperature, which have an effect on the physical condition of the carbon dioxide, play a significant role in determining the capacity of geological media to store carbon dioxide. It is possible for carbon dioxide to assume a state in which it has the density of a liquid but the viscosity of a gas when it is subjected to supercritical conditions, which normally occur at pressures greater than 7.38 megapascals and temperatures more than 31.1 °C [22–24]. Because of this condition, carbon dioxide is able to fill a greater amount of pore space inside the reservoir while yet retaining a high degree of mobility [22, 23]. This state is ideal for storage. It is possible to improve the storage efficiency of carbon dioxide as well as its capacity to infiltrate the reservoir by ensuring that it remains in this supercritical phase. For the purpose of forecasting how carbon dioxide behaves in subterranean formations, it is essential to have a solid understanding of the phase behaviour of carbon dioxide under different pressure and temperature conditions. In order to ensure that carbon dioxide (CO_2) stays in its supercritical state and, as a result, maximises the storage capacity of geological formations, phase diagrams (Fig. 2.2), which depict these conditions, are extremely useful [4, 22–24].

The quantity of space that is available to store carbon dioxide is directly influenced by porosity, which is a significant factor in determining the storage capacity of geological formations. Porosity, on the other hand, tends to decrease with increasing

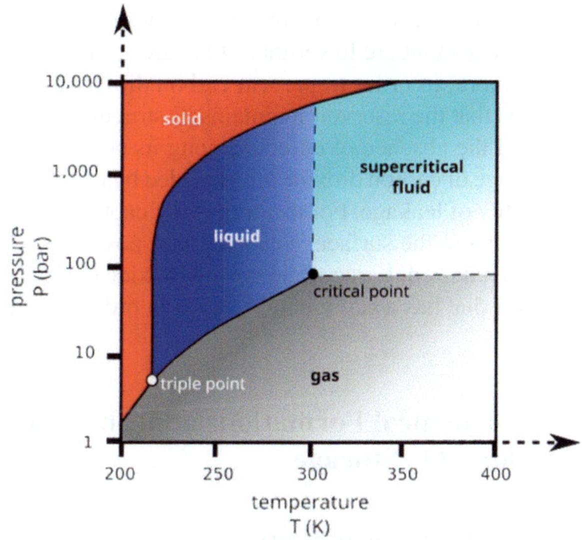

Fig. 2.2 Phase diagram of CO_2

depth as a result of natural processes such as compaction and cementation, which cause the rock to become thicker and lower the number of pore spaces that are available [25]. The variety in the physical structure and composition of rock strata that are created by diverse depositional conditions is referred to as stratigraphic heterogeneity. This is another important aspect that has a significant impact on the efficiency of storage. It is common for the transport of buoyant carbon dioxide to be focused along certain pathways in formations where the rock is relatively homogeneous. This results in a significant portion of the rock volume being abandoned. Consequently, this may lead to a reduction in the total storage capacity [24–26]. On the other hand, carbon dioxide (CO_2) can be dispersed more uniformly throughout a wider variety of flow routes in heterogeneous formations. This is because the rock in these formations has variances in its structure and permeability when compared to other formations. It is possible to store a bigger volume of carbon dioxide as a result of the diversity in flow channels, which contributes to an overall improvement in the efficiency of the storage process. Although a high porosity is necessary for the storage potential of the rock, the structural complexity of the rock also plays a vital role in optimising the quantity of carbon dioxide that can be stored. When compared to homogeneous formations, heterogeneous formations, which have a variety of flow patterns, are often more effective at accommodating higher volumes of carbon dioxide; this is according to research [25]. Sites that are capable of effectively storing huge quantities of carbon dioxide are required for efficient CO_2 storage. When it comes to estimating the storage capacity and efficiency of the site, factors such as the depth of the reservoir and the total organic carbon (TOC) content play a significant impact. In addition, conditions of high pressure and high temperature help to the maintenance of CO_2 in a supercritical state, which is more effective for storage [8, 24]. It is the capacity of a CO_2 storage site to contain CO_2 over an extended period of time with a minimal amount of leakage that determines the reliability of the site. Sites that have leakage rates that are lower than 0.1% are regarded as being extremely dependable. Furthermore, in order to prevent carbon dioxide from escaping to the surface, it is essential that the reservoir maintains its structural integrity, which includes its thickness and the absence of defects among its components. More stable conditions for the storage of carbon dioxide are provided by deeper reservoirs, which also lower the possibility of leakage. For the purpose of capturing carbon dioxide and preventing it from rising to the surface, a caprock that possesses a high sealing capacity, such as shale, is required. Large-scale carbon dioxide (CO_2) sequestration projects require a reservoir that has a sufficient thickness in order to guarantee a high storage volume.

2.4 Geological Formations and Their Screening Criteria for CO_2 Storage

2.4.1 Shale Formations

When selecting shale formations for the purpose of CO_2 storage, it is necessary to take into consideration a number of characteristics in order to guarantee efficient and safe sequestration [10]. To ensure that the conditions of pressure and temperature are maintained at an appropriate level, the depth of the shale formation should be greater than 800 m. To ensure that there is sufficient capacity for storage, the thickness should be greater than fifty meters. In order to maintain the CO_2 in a supercritical state, the temperature conditions must be higher than 35 °C, and the pressure must be more than 7.5 megapascals. The permeability of a material must be more than 0.1 millidarcy (md) in order to permit CO_2 injection, whereas the porosity of the material might range from 1 to 10%, which shows its potential to store CO_2. It is optimal to have a formation thickness that is greater than one hundred meters overall. In order to guarantee effective CO_2 displacement and storage, it is essential that oil-bearing shales have an oil density that exceeds 0.9, a viscosity that falls within the range of 10–1000 mPa, and an oil saturation that falls between 0.3 and 0.7 [11], although this is not always the case.

2.4.2 Sandstone and Carbonate Formations

It is suggested that the depth be above 450 m for sandstone and carbonate formations, as this is the optimal depth for maintaining the required pressure conditions. The thickness of the reservoir should be less than forty meters, and the temperature should be between sixty and one hundred and twenty degrees Celsius. This will ensure that the carbon dioxide remains in a dense phase. It is necessary for the pressure to be higher than 7.5 MPa. The porosity should be larger than three percent, which indicates that there is sufficient area for storage, and the permeability should be greater than five percent in order to enable efficient CO_2 injection. It is preferable to have a thickness of more than ten meters in order to provide sufficient storage capacity. When it comes to reservoirs that include oil, it is essential to ensure that the oil density is less than 0.88, the oil viscosity is below 10 mPa·s, and the oil saturation is greater than 0.3. This is necessary in order to guarantee efficient CO_2 displacement and more efficient oil recovery. It is possible to identify geological formations that are ideal for carbon dioxide sequestration with the use of these screening criteria, which guarantees the safety, efficiency, and reliability of the places for storage [8, 11, 13, 21].

2.4.3 Coal

For the purposes of determining the screening criteria for CO_2 storage in coal, the mechanisms largely involve the processes of sorption and diffusion [27]. Carbon dioxide is said to diffuse as it travels via bigger pores, fractures, and cleats in coal formations. These pores and fractures often have a size that is greater than 30 nm. Adsorption onto the internal surfaces of the coal, absorption into its molecular structure, the existence of free gas in fractures, cleats and voids, and dissolving in groundwater are all examples of sorption mechanisms [27]. Adsorption onto the internal surfaces of the coal is another example of sorption mechanisms. For the purpose of facilitating efficient storage and retention, it is necessary to have sufficient permeability across the pores and cracks in coal seams in order to achieve successful injection of carbon dioxide (Table 2.1).

2.4.4 Environmental Based Site Screening

The process of analysing prospective places for carbon dioxide geological storage with the goal of minimising the negative environmental effects is referred to as environmental concern-based site screen from the site screening perspective. This strategy incorporates a number of criteria in order to guarantee that the sites that are chosen not only fulfil the criteria for technical and economic requirements, but also comply to the standards for safeguarding the environment and sustainability. A few essential variables to consider are:

(a) Evaluating the potential effects of carbon dioxide storage on local ecosystems, including the flora and fauna, is the first step in the ecosystem impacts process. In this context, the assessment of the danger of carbon dioxide (CO_2) leakage to the surface, as well as its influence on soil and plants, and any potential changes to environmental habitats for local wildlife, is included.

Table 2.1 Screening criteria for CO_2 sequestration in different geological media [12, 13, 22, 27]

Geological formation	Screening criteria	Example
Depleted oil and gas reservoirs	• High porosity and permeability	North Sea Sleipner Project, Norway
	• Adequate depth	Weyburn-Midale CO_2 Project, Canada
	• Structural integrity	
	• Existing infrastructure	
Saline formations	• Large storage capacity	Mount Simon Sandstone, USA
	• High salinity levels	Gorgon CO_2 Injection Project, Australia
	• High porosity and permeability	
	• Good caprock integrity	
Deep coal beds	• High sorption capacity	San Juan Basin, USA
	• Adequate permeability	Qinshui Basin, China
	• Enhanced Coal Bed Methane (ECBM) potential	
Basalt formations	• Reactive with CO_2 to form stable carbonates	CarbFix Project, Iceland
	• High permeability and porosity	Columbia River Basalt Group, USA
	• Favorable mineral composition	
Shale formations	• Low permeability (acting as caprock)	Marcellus Shale, USA
	• Adequate depth	Eagle Ford Shale, USA
	• Good structural integrity	

(b) Seismic Activity: Taking into consideration the possibility of experiencing seismic activity as a result of the injection of carbon dioxide. In order to accomplish this, it is necessary to conduct an analysis of the seismic history of the region, as well as the presence of fault lines and the possibility of reactivation of fractures and faults as a result of increased pressure caused by CO_2 injection.

(c) Human Health and Safety: Ensuring the safety of populations in the surrounding area by assessing the possibility of carbon dioxide leakage and the effect that it has on the quality of the air. The modelling of the dispersion of carbon dioxide in the case of a leak and the evaluation of the risk to human health are both included in this.

(d) Compliance with legislation: Making certain that the process of site selection is in accordance with environmental legislation and guidelines at the local, national, and international levels to ensure compliance. In order to accomplish this, it is necessary to be current on the ever-changing regulations and to include them into the screening process.

(e) Long-term Monitoring and Management: The process of developing comprehensive monitoring and management plans through which any potential environmental consequences can be identified and mitigated over the course of a longer period of time. The implementation of continuous monitoring systems for CO_2 levels, pressure changes, and potential leakage channels is included in this (Table 2.2).

The thresholds used for the pre-screening of CO_2 storage sites in the Netherlands (Table 2.3), as outlined by Ramírez et al. [26, 28, 29].

The depth of the storage formation, the requirement for extra infrastructure, and the operational expenses are all factors that might have an impact on the charges associated with storage. During the selection process, priority is given to locations that have lower storage costs because these locations are significantly more economically appealing. They minimise transportation costs and logistical obstacles, which makes the whole storage project more practical [26, 28, 29]. Sites that are closer to CO_2 emission sources and existing infrastructure are chosen because they lower the amount of money spent on transportation.

Table 2.2 CCUS and environmental impact [1, 2, 25, 28, 29]

Environmental element	Negative impact	Fatal or serious impact
Groundwater and surface water	• Low acidity caused at 0.2–2% concentration of CO_2, causing minor changes • Moderate acidity and corrosion at CO_2 concentration >2%	• Significant increase in acidity at CO_2 concentration >6% • Enhanced corrosion and loss of irrigation effectiveness
Vegetation	• CO_2 concentration >5% causes detrimental effects on plant health and yield • Serious impact on vegetation at 5–30% concentration	• Long-term CO_2 concentration >20% in soil gas results in dead zones where no plants survive • Concentrations >30% are fatal to plant life
Human health	• CO_2 concentration of 1–3% causes shortness of breath, headaches, and sweating, with physiological adaptation • Concentration of 3–5% leads to shortness of breath, high blood pressure, and discomfort • Concentration >5% causes physical and mental harm and loss of consciousness	• CO_2 concentration >10% causes severe symptoms, rapid loss of consciousness, coma, or death • Prolonged exposure to CO_2 concentration of 25–30% leads to respiratory arrest and death

Table 2.3 .

Parameter	Threshold
Capacity	≥4 Mt for gas/oil and ≥2 Mt for aquifer
Thickness reservoir	>10 m
Depth top reservoir	≥800 m
Reservoir porosity	Aquifers: >10%
Reservoir permeability	Aquifers: an expected permeability of 200 mD or more
Thickness seal	≥10 m. Both simple seals as well as complex seal have been taken into account
Seal composition	Salt, anhydrites, shale or claystone
Reservoir composition	Aquifers: sandstone, hydrocarbon fields: limestone, sandstone, siltstone, carbonates
Initial pressure	Overpressure excluded
Salt domes	Relevant for aquifers. Traps located alongside/near salt domes/walls have been excluded because there is a high risk of salt cementation

2.5 Conclusion

There are a total of almost 764 basins in the world. There are a variety of geological formations, each of which has distinctive features and mechanisms of trapping. Currently, a large number of industries, institutes, and research organisations are focussing their efforts on the sequestration of carbon dioxide in geological formations. With the goal to achieve better sequestration in geological formations, it was necessary to do extensive research on the reservoir's physical, chemical, and mechanical properties. For the purpose of further research, the fracture study, leakage study, pressure study, and wettability study of the reservoir are among the most significant characteristics to investigate.

References

1. Wennersten R, Sun Q, Li H (2015) The future potential for carbon capture and storage in climate change mitigation—an overview from perspectives of technology, economy and risk. J Clean Prod 103:724–736
2. Szulczewski ML, MacMinn CW, Herzog HJ, Juanes R (2012) Lifetime of carbon capture and storage as a climate-change mitigation technology. Proc Natl Acad Sci 109(14):5185–5189
3. Saadatpoor E, Bryant SL, Sepehrnoori K (2010) New trapping mechanism in carbon sequestration. Transp Porous Media 82:3–17

4. Aminu MD, Nabavi SA, Rochelle CA, Manovic V (2017) A review of developments in carbon dioxide storage. Appl Energy 208:1389–1419
5. Zhang D, Song J (2014) Mechanisms for geological carbon sequestration. Procedia IUTAm 10:319–327
6. Huppert HE, Neufeld JA (2014) The fluid mechanics of carbon dioxide sequestration. Annu Rev Fluid Mech 46(1):255–272
7. Gunter WD, Bachu S, Benson S (2004) The role of hydrogeological and geochemical trapping in sedimentary basins for secure geological storage of carbon dioxide. Geol Soc, London, Spec Publ 233(1):129–145
8. Ramadhan R, Tapanya C, Akamine T, Leelasukseree C, Tangparitkul S (2024) CO_2 trapping dynamics in tight sandstone: insights into trapping mechanisms in Mae Moh's reservoir. J Environ Manage 370:122442
9. Mitchell AC, Dideriksen K, Spangler LH, Cunningham AB, Gerlach R (2010) Microbially enhanced carbon capture and storage by mineral-trapping and solubility-trapping. Environ Sci Technol 44(13):5270–5276
10. Boruah A, Rasheed A, Mendhe VA, Ganapathi S (2019) Specific surface area and pore size distribution in gas shales of Raniganj Basin, India. J Pet Explor Prod Technol 9:1041–1050
11. Boruah A, Ganapathi S (2015) Microstructure and pore system analysis of Barren Measures shale of Raniganj field, India. J Nat Gas Sci Eng 26:427–437
12. Hazarika S, Boruah A (2022) Supercritical CO_2 (SCO_2) as alternative to water for shale reservoir fracturing. Mater Today: Proc 50:1754–1757
13. Hazarika S, Boruah A, Saraf S (2023) Modeling and simulation study of CO_2 fracturing technique for shale gas productivity: a case study (India). Arab J Geosci 16(7):408
14. Boruah A, Prusty BK (2023) Influence of mineralogy, organic matter and thermal maturity on the diagenetic evolution of pore systems in the shales of Cambay and Raniganj Basin, India. J Geol Soc India 99(11):1577–1585
15. Steele-MacInnis M, Capobianco RM, Dilmore R, Goodman A, Guthrie G, Rimstidt JD, Bodnar RJ (2013) Volumetrics of CO_2 storage in deep saline formations. Environ Sci Technol 47(1):79–86
16. Rasmusson M (2018) Residual and solubility trapping during geological CO_2 storage: numerical and experimental studies. Doctoral dissertation, Acta Universitatis Upsaliensis
17. Suekane T, Nobuso T, Hirai S, Kiyota M (2008) Geological storage of carbon dioxide by residual gas and solubility trapping. Int J Greenhouse Gas Control 2(1):58–64
18. Burnside NM, Naylor M (2014) Review and implications of relative permeability of CO_2/brine systems and residual trapping of CO_2. Int J Greenhouse Gas Control 23:1–11
19. Khandoozi S, Hazlett R, Fustic M (2023) A critical review of CO_2 mineral trapping in sedimentary reservoirs–from theory to application: pertinent parameters, acceleration methods and evaluation workflow. Earth-Sci Rev 104515
20. Bachu S, Gunter WD, Perkins EH (1994) Aquifer disposal of CO_2: hydrodynamic and mineral trapping. Energy Convers Manage 35(4):269–279
21. Xu T, Apps JA, Pruess K (2004) Numerical simulation of CO_2 disposal by mineral trapping in deep aquifers. Appl Geochem 19(6):917–936
22. Bachu S (2000) Sequestration of CO_2 in geological media: criteria and approach for site selection in response to climate change. Energy Convers Manage 41(9):953–970
23. Nassabeh M, You Z, Keshavarz A, Iglauer S (2024) Sub-surface geospatial intelligence in carbon capture, utilization and storage: a machine learning approach for offshore storage site selection. Energy 305:132086
24. Luboń K (2021) Influence of injection well location on CO_2 geological storage efficiency. Energies 14(24):8604
25. Solomon S (2007) Carbon dioxide storage: geological security and environmental issues—case study on the sleipner gas field in Norway. Bellona report 128
26. Ramírez A, Hagedoorn S, Kramers L, Wildenborg T, Hendriks C (2010) Screening CO_2 storage options in the Netherlands. Int J Greenhouse Gas Control 4(2):367–380

27. Singh S, Boruah A, Devaraju J (2023) CO_2 sequestration and CH_4 extraction from unmineable coal seams in Singrauli coalfield. Int J Coal Prep Utilization 1–18
28. Welkenhuysen K, Ramírez A, Swennen R, Piessens K (2013) Strategy for ranking potential CO_2 storage reservoirs: a case study for Belgium. Int J Greenhouse Gas Control 17:431–449
29. Raza A, Rezaee R, Gholami R, Bing CH, Nagarajan R, Hamid MA (2016) A screening criterion for selection of suitable CO_2 storage sites. J Nat Gas Sci Eng 28:317–327

Chapter 3
Carbon Dioxide Injection and Gas Recovery from the Shale Formation: Experimental and Simulation Study

Abstract CO_2 can be permanently store in the shale reservoirs and also can be utilised for enhancing the gas extraction from the shale. This chapter focuses on shale rock characterization in terms of mineral composition, porosity, permeability, pore size distribution, and grain density, and also simulation for CO_2 storage estimation. The current study is based on both field study, laboratory investigation, and reservoir modelling. CO_2 sequestration potential in the Chalbalpur-Mahishmura regions of the Raniganj Field, India was assessed with the help of simulation techniques. The studied shales show the porosity levels between 0.02% and 1.21%, permeability from 0.10 to 1.67 millidarcies (mD), and grain density ranging from 2.40 to 3.09 g per cubic centimetre (g/cc). Simulation results the total cumulative volume of CO_2 storage capacity 721,000m^3, with around 1296.3 tons of CO_2 potentially sequestered over 300 to 400 days within the shale reservoir.

Keywords Shale formation · CO_2 injection · Simulation · Porosity · CO_2 storage

3.1 Introduction

Increased CO_2 emission is a primary factor in environmental degradation and global warming [1–3]. Reducing CO_2 from the atmosphere is vital for addressing climate change. One promising method is injecting and storing CO_2 in geological formations, like shale reservoirs. High porosity and low permeability of shales makes it suitable for CO_2 storage. However, a deep understanding of the physical, chemical and reservoir properties are needed. Key reservoir properties i.e. porosity, pore size distribution, grain density, and permeability needs to be assessed and feasibility of CO_2 injection in a shale reservoir should be performed before injection [1–4]. These properties are important for assessing the storage capacity and viability of a proposed injection site. Research has shown that the pore structure of shale formations plays a critical role in CO_2 sequestration, as variations in pore sizes can affect gas injection. An increase in mesopores and micropores, for instance, can improve CO_2 storage capacity, with pore sizes between 0.35 and 300 nm being particularly effective for

A. Boruah, *CO_2 Geosequestration: Capturing Carbon for a Sustainable Future*, Springer Climate, https://doi.org/10.1007/978-3-031-81021-3_3

gas adsorption [1–5]. Shale reservoirs, however, tend to have lower permeability compared to other formations, such as coal beds [6, 7]. Therefore, CO_2 injection into shale can help with gas recovery and sequestration, the complexity of shale's geological properties and its interactions with CO_2 must be fully understood to optimize these processes.

The injection of carbon dioxide (CO_2) into shale formations for enhanced gas recovery and long-term storage is influenced by several critical factors. Key geological properties of the shale reservoir, such as porosity, permeability, and mineral composition, play a fundamental role in the injection and recovery processes [8–11]. Porosity dictates the amount of space available for CO_2 storage, while permeability controls the ease with which CO_2 can move through the formation. However, shale formations typically have low porosity and permeability, which can reduce the efficiency of CO_2 injection and distribution within the reservoir. Understanding the mineral composition of the shale is also important, as it affects how CO_2 interacts with the rock matrix. For instance, shales with higher clay content may absorb more CO_2, impacting both gas recovery and storage potential. Subsurface pressure and temperature conditions are crucial for determining the phase behavior of CO_2 [10, 11]. CO_2 can be injected in its supercritical state at higher depth, where it exhibits the properties of both a liquid and a gas. This supercritical CO_2 can effectively permeate the reservoir and occupy pores, enhancing both storage capacity and gas displacement. Maintaining the appropriate pressure during the injection process is essential to avoid fracturing the formation, which could compromise both the integrity of the reservoir and the effectiveness of gas recovery [5, 10, 11]. The interaction between the injected CO_2 and the existing natural gas within the shale formation is another key factor. CO_2 displaces methane (CH_4) via adsorption and desorption process. In this mechanism, CO_2 molecules adsorb onto the surface of the shale, displacing methane that was previously adsorbed onto the rock surface, thereby enhancing gas recovery. The effectiveness of this process depends on the adsorption capacity of the shale, which is influenced by factors such as pore structure, pressure, and the chemical composition of the shale. The rate at which CO_2 is injected into the shale formation significantly impacts both the efficiency of gas recovery and the long-term storage of CO_2 [10–12]. A higher injection rate may increase the pressure within the formation too quickly, leading to potential fracturing or bypassing of certain areas of the reservoir. Conversely, slow injection rates are generally economically not viable over time. The timing of CO_2 injection, particularly in relation to the reservoir's production history, also plays a role. Injecting CO_2 after significant gas depletion may result in better recovery, as more pore space becomes available for CO_2 storage and gas displacement [10, 11]. The physical stress and strain on the shale formation during CO_2 injection can alter the reservoir's characteristics. Over time, CO_2 injection may lead to changes in the reservoir's permeability and porosity, as well as the creation of fractures. These geomechanical effects must be carefully managed to ensure that CO_2 remains securely trapped within the formation and that gas recovery is optimized without damaging the integrity of the reservoir [10, 11]. The movement of CO_2 and natural gas within the shale formation is governed by the fluid dynamics of the system. Due to the heterogeneity of shale formations, CO_2

may follow preferential flow paths, bypassing certain regions of the reservoir. This uneven distribution can result in suboptimal gas recovery and CO_2 storage. Understanding the flow behavior within the shale and implementing strategies to ensure more uniform distribution of CO_2 are crucial for maximizing both recovery and storage efficiency [8–11].

The effective range of the pore diameters, pore percentage and the permeability are essential for the injection of CO_2 in shale formation [12–15] [22]. The understanding of the effectiveness of these factors is depend on the cumulative volume, and so it needs to forecast the cumulative volume. The cumulative volume estimation can be carried out using specific reservoir modelling and simulation techniques. Reservoir modelling and simulation techniques are the mathematical models for the prediction of CO_2 injection potential of any reservoir [16–19]. In the first step, the numerical modeling is created based on the 2D or 3D model of the reservoir. The specification of the developed reservoir model portraits as the actual reservoir environment. In the second step, the simulated model of the reservoir needs to analyzed based on the in-situ condition, rock characteristic parameters and boundary conditions of the reservoir. The reservoir simulation helps is understanding the real performance of the production well with respect to a specific time period. In this study, the reservoir simulation was carried out for the prediction of production rate and cumulative gas volume over the life of the production well [8–13, 20]. Present study also focuses on the sensitivity analysis and history matching along with the recourse estimation [12].

This chapter illustrates the reservoir parameters of Barren Measures shale in Raniganj formation, India in terms of its porosity, pore size distribution, grain density, and permeability. Statistical methods were applied to established the mutual correlation between different parameters to validate CO_2 injection efficiency in Raniganj Field. CMG-GEM was applied to perform numerical analysis using finite difference simulators to estimate cumulative gas volume based on the reservoir data.

3.2 Geology of the Study Area

The study area is situated between the coordinates 23°46′00″ N to 23°43′00″N in latitude and 86°52′00″E to 86°55′30″E in longitude, as depicted on the Survey of India toposheet numbers 73I/13 and 73I/14. Fieldwork was conducted along the exposed sections of the Barren Measures Formation in the Chalbalpur-Mahishmura regions, which are part of the Raniganj Formation in India.

In this region, the Permian shales consist of three primary formations: Barakar, Barren Measures, and Raniganj. The Barakar Formation includes a mix of sandstone, shale, and coal. The Barren Measures Formation, which lies above the Barakar Formation, is primarily composed of shale and sandstone. Finally, the Raniganj Formation, which is situated higher in the stratigraphic sequence, is predominantly made up of coal, shale, and sandstone [14] (Fig. 3.1).

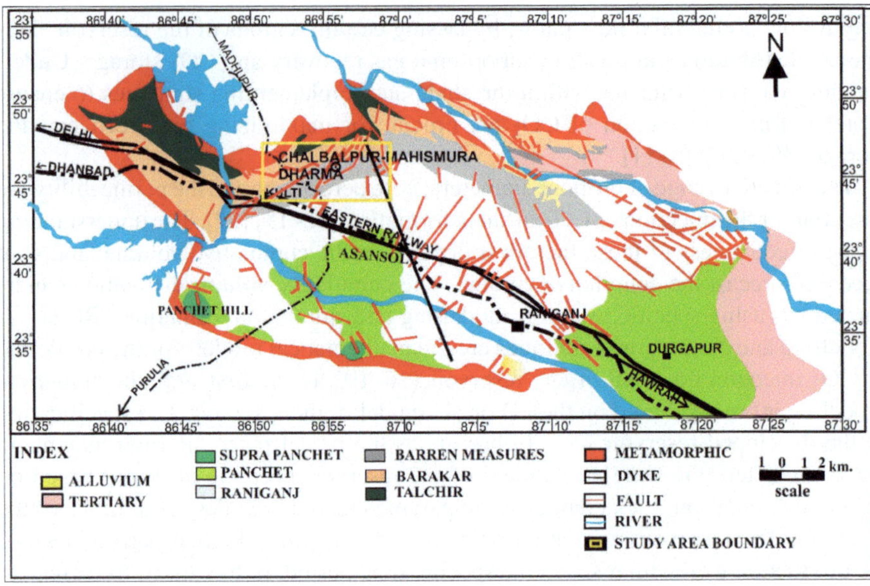

Fig. 3.1 Geological location of Raniganj Formation India showing Chalbalpur-Mahishmura areas [14]

3.3 Methodology

The shale samples were collected from the Raniganj Field to carryout extensive loboratory studies. Shale core samples were analysed using Ultrapore-300 helium porosimeter to determine the porosity and grain density using on Boyles law [14]. The cylindrical core samples were prepared for the permeability using Ultraperm 400. This works on the basis of Darcy's law and determines the steady state flow rate through the core at specific pressure gradient. This apparatus acquires data as real time graphical display with mass flow to provide more accurate, valid and precise permeability data. It also shows flow slippage effect to derive Klinkenberg corrected permeability [15].

The shale samples were assesses using Brunauer–Emmett–Teller (BET) adsorption test to determine the surface area, pore size distribution and adsorbed gas content. Nitrogen gas was used in this test to carried out multilayer adsorption/desorption. The Micromeritics ASAP 2010 BET Tester was utilized for the determination of above parameters [15].

3.4 Results and Discussion

Shale samples from four boreholes at varying depths in the study area were analyzed to assess key reservoir properties, including porosity, grain density, and permeability. These properties—along with mineralogy—are critical in determining the gas storage capacity, flow behavior, and hydraulic fracturing potential of shale gas reservoirs. The analysis revealed that as depth increased, porosity decreased while grain density increased, a trend attributed to compaction and geological factors. Porosity values ranged from 0.014% to 1.347%, with grain density between 2.4 and 2.57 g/cc (Table 3.1). Air permeability values ranged from 0.1 to 0.167 mD, though some samples showed negligible permeability due to the limitations of conventional testing methods for tight shale, requiring more advanced techniques for accurate measurement. Key factors affecting gas reserve calculations include the areal extent of the reservoir, porosity, pay thickness, and total gas volume (free and adsorbed gas). Organic matter and mineral content play vital roles in determining reservoir continuity, effective porosity, and brittleness for hydraulic fracturing. To better understand porosity, theoretical net porosity values from core analysis were compared with pore size distributions generated through BET techniques, helping to estimate the storage potential of the shale reservoir. The study reported that the Barren Measures shale specific surface area ranges from 7.67 to 16.21 m^2/g. BJH adsorption reveals the pore size diameter range from 6.09 nm (60.9A°) to 34.56 nm (345.6A°). It reveals the presence of mesopores and the changes in the pore volume of Barren Measures due to the microstructure heterogeneity in terms of mineral content and volatile matters and diagenesis changes. The studied samples show that the surface area and volume decrease with an increasing the diameter of the pore. It also indicates the reduction of available micro and mesopore volumes.

3.4.1 Results of Reservoir Simulation

The reservoir simulation was conducted using experimentally derived data, with some parametric values estimated based on a literature review. The parameters utilized in the simulation are listed in Table 3.2. A 20 × 20 × 6 Cartesian grid was employed for model development, as shown in Fig. 3.2. CO_2 injection was simulated using an injection well within the perforated reservoir, as illustrated in Fig. 3.2.

The total gas in place and gas saturation were determined for the developed model based on various reservoir parameters outlined in Table 3.2. The simulated results are shown in Figs. 3.2, 3.3, 3.4, 3.5, 3.6, 3.7, 3.8, 3.9, 3.10, 3.11, 3.12, 3.13, 3.14. The maximum gas in place ranged from 349 to 1385 m^3 for borehole #1 to borehole #4. Differences in gas volumes between the boreholes were due to variations in porosity and permeability. Boreholes #2 and #4, having higher porosity compared to the others, resulted in more repository sites and thus higher gas volumes. Borehole #2, with a permeability of 0.148 mD—the highest among all—also had the greatest

Table 3.1 Results of helium porosimetry, grain density and air permeability [15]

Boreholes	Depth (m)	Porosity (%)	Grain density (g/cc)	BET surface area (m^2/g)	Adsorption Avg. pore diameter (4 V/A by BET)
#1	150	1.104	2.515	14.19	71.20
	155	0.9	2.406		
	170	0.95	2.4		
	185	0.546	2.417		
	195	0.474	2.415		
#2	90	0.959	2.567	15.11	54.83
	96	1.055	2.57		
	100	1.347	2.521		
	115	0.014	2.537		
	150	1.02	2.526		
	155	0.8	2.407		
#3	110	0.972	2.5	11.75	59.69
	115	1.211	2.552		
	120	0.366	2.464		
	125	0.129	2.458		
	154	0.02	3.089		
#4	305	0.08	2.45	13.27	56.04
	315	0.09	2.482		
	320	0	3.09		

*NC = Not calculated

gas in place (Fig. 4a). Gas saturation across the boreholes ranged from 60 to 90% near the injection well, with distribution on both sides of the fractured reservoir, indicating effective CO_2 injection (Figs. 3.3, 3.4, 3.5, 3.6, 3.7, 3.8).

The performance of the injection well was evaluated for all boreholes using the data provided in Table 3.2. The simulation was conducted over 400 days with a surface gas injection rate of 1000 m^3/day (Figs. 3.9, 3.10). A decline in gas flow rate was observed after 140 days, reaching zero by day 180 (Fig. 3.10), suggesting reservoir saturation after 180 days of CO_2 injection in Borehole #1. The well bottom hole pressure increased to 50 MPa by day 180, with a maximum cumulative gas volume of 1.5×10^5 m^3 (Fig. 3.7). Boreholes #2, #3, and #4 exhibited gas rate declines at 250, 70, and 220 days, respectively (Figs. 3.8, 3.9, 3.10). For Boreholes #2 and #4, the surface gas rate dropped to zero at 300 days, while Borehole #3 reached this point at 80 days. The maximum bottom hole pressure reached 50 MPa for all boreholes. The cumulative gas volumes were 2.7×10^5 m^3 for Borehole #2, 0.61×10^5 m^3 for Borehole #3, and 2.4×10^5 m^3 for Borehole #4 (Figs. 3.8, 3.9, 3.10, 3.11, 3.12). Variations in injection rate and cumulative gas volume were attributed to

Table 3.2 List of important parameters used for the reservoir simulation [15]

Reservoir parameters	Values
Grid type	Cartesian
Grid	20 × 20 × 6
Total block	2400
Porosity type	Single
Pinchout thickness	0.0002
Grid top	155–305 m
Grid thickness	1 m
Porosity	0.02–0.8%
Permeability I Permeability J Permeability K	0.1–0.148 mD
Rock compressibility	4.5×10^{-7} 1/kPa
Reference pressure	4185–8235 kPa
Depth	155–305 m
Reservoir temperature	35 ºC
Model for equation of state	Peng-Robinson
Density	2407–3092 kg/m^3
Surface gas rate	1000 m^3/d
Bottom hole pressure	50,000 kPa
Injected fluid	CO_2
Total time of simulation	400 Days

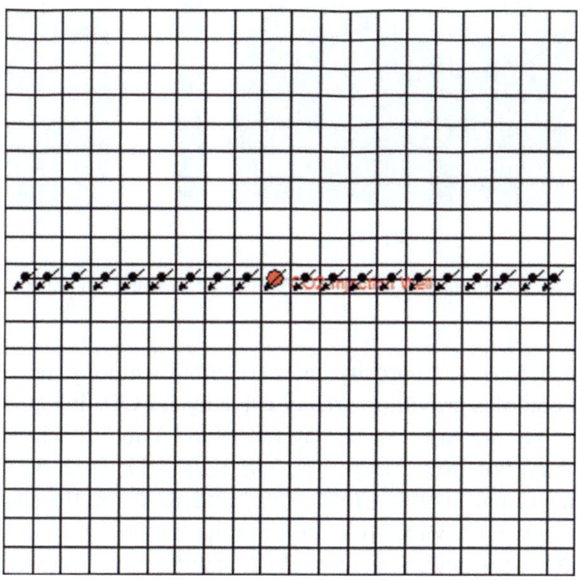

Fig. 3.2 Reservoir model along with the injection well

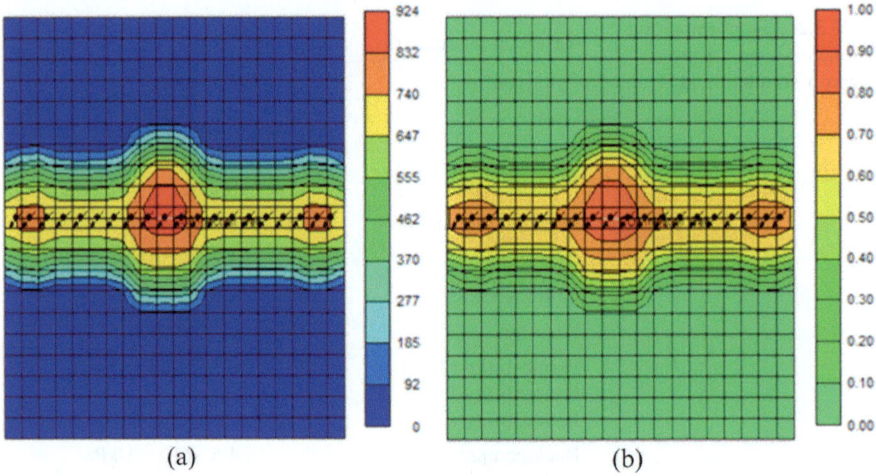

Fig. 3.3 Simulation result of borehole#1 with the contour fill of **a** gas in place, **b** gas saturation

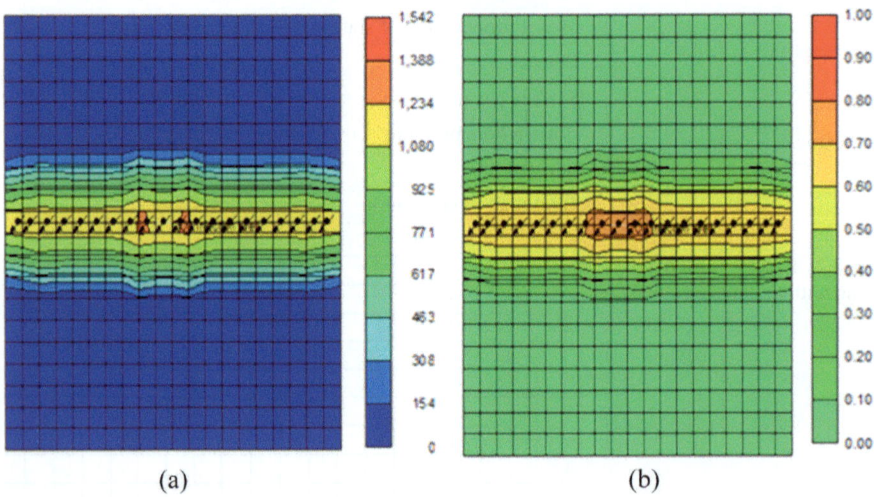

Fig. 3.4 Simulation result of borehole #2 showing contour fill of **a** gas in place, **b** gas saturation

differences in porosity and permeability among the boreholes, with higher porosity boreholes providing greater capacity for CO_2 injection.

An increase in CO_2 saturation and a decrease in water saturation were observed in all boreholes (Figs. 3.8, 3.9, 3.10, 3.11, 3.12, 3.14). CO_2 saturation levels were measured at 90%, 75%, 65%, and 62% for Boreholes 1 through 4, respectively (Figs. 3.12, 3.13, 3.14). In Borehole 1, CO_2 saturation stabilized after approximately 200 days, while in Borehole 2, stabilization occurred around 300 days (Figs. 3.11, 3.12, 3.13, 3.14). For Boreholes 3 and 4, gas saturation continued beyond 300 days

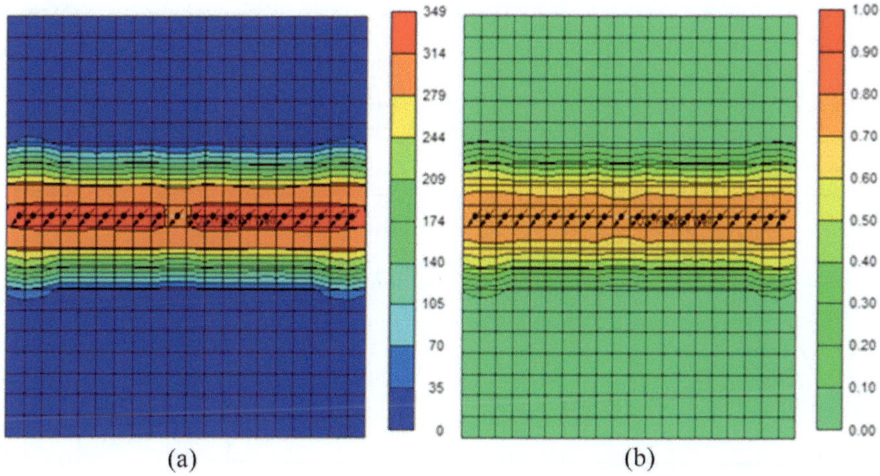

Fig. 3.5 Simulation result of borehole #3; contour fill of **a** gas in place, **b** gas saturation

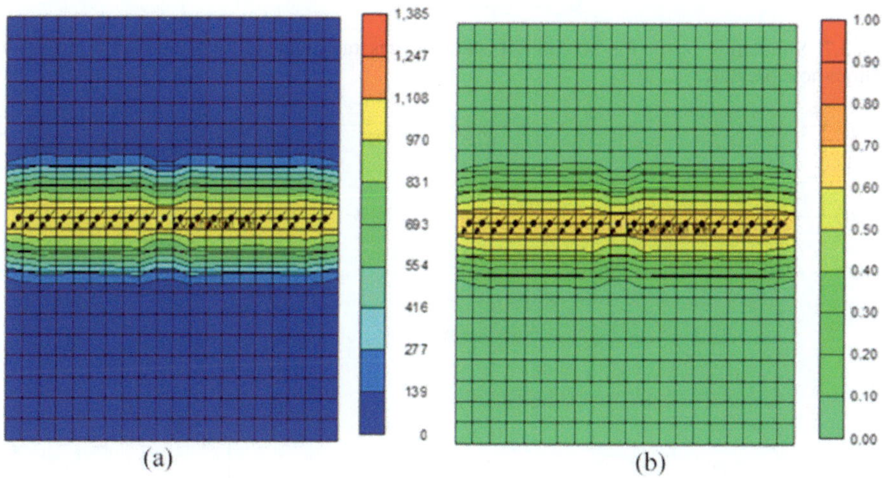

Fig. 3.6 Simulation result of borehole #4; contour fill of **a** gas in place, **b** gas saturation

(Figs. 3.13, 3.14). These results suggest the potential for additional CO_2 injection in Boreholes 2 through 4.

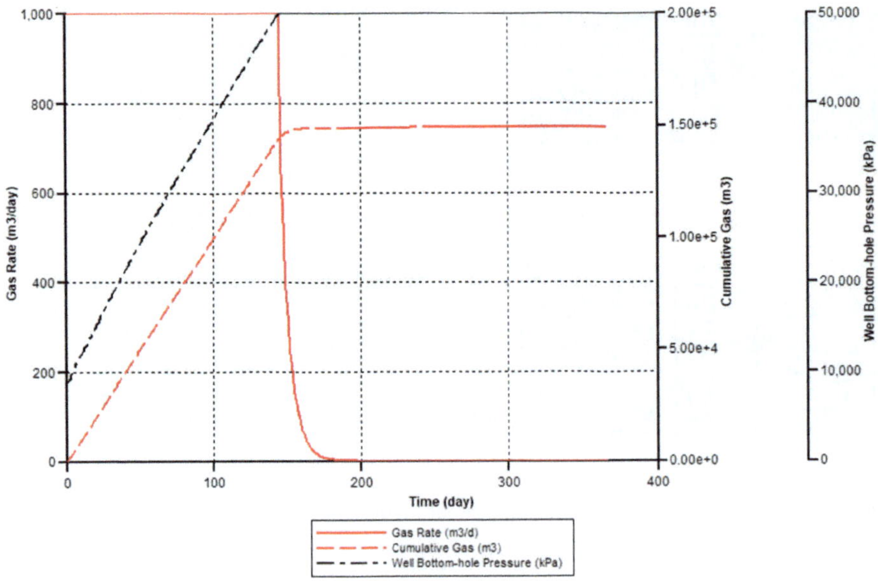

Fig. 3.7 Simulation result for borehole#1 showing gas rate, cumulative gas volume and well bottom-hole pressure

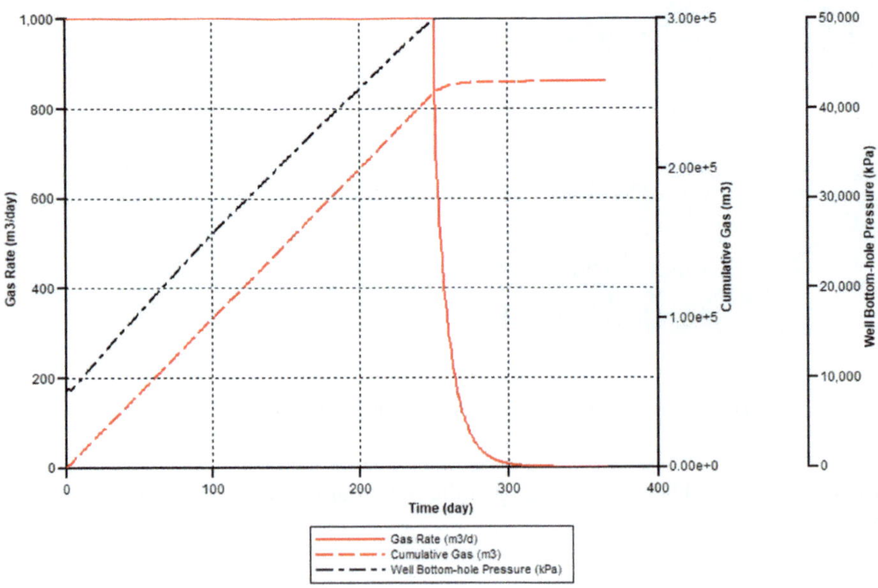

Fig. 3.8 Simulation result for borehole#2 showing gas rate, cumulative gas volume and well bottom-hole pressure

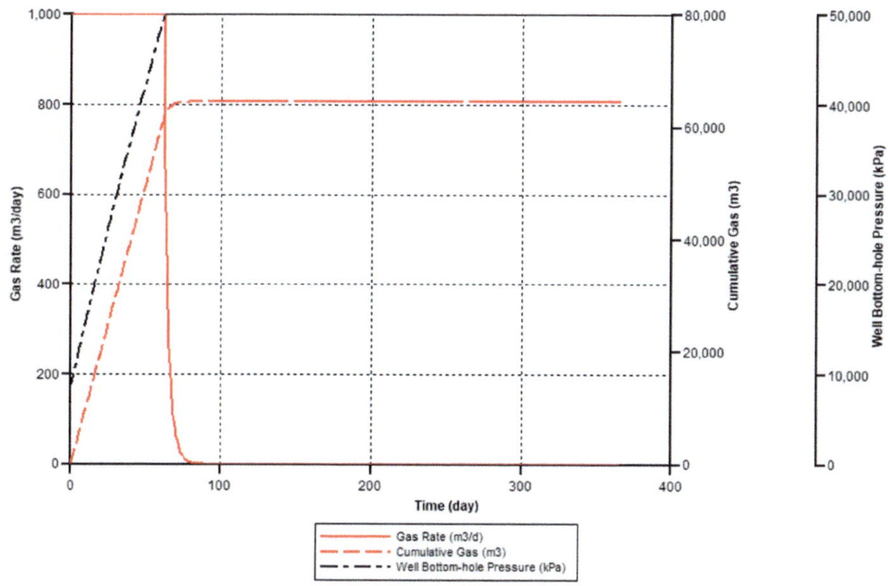

Fig. 3.9 Simulation result for borehole#3 showing gas rate, cumulative gas volume and well bottom-hole pressure

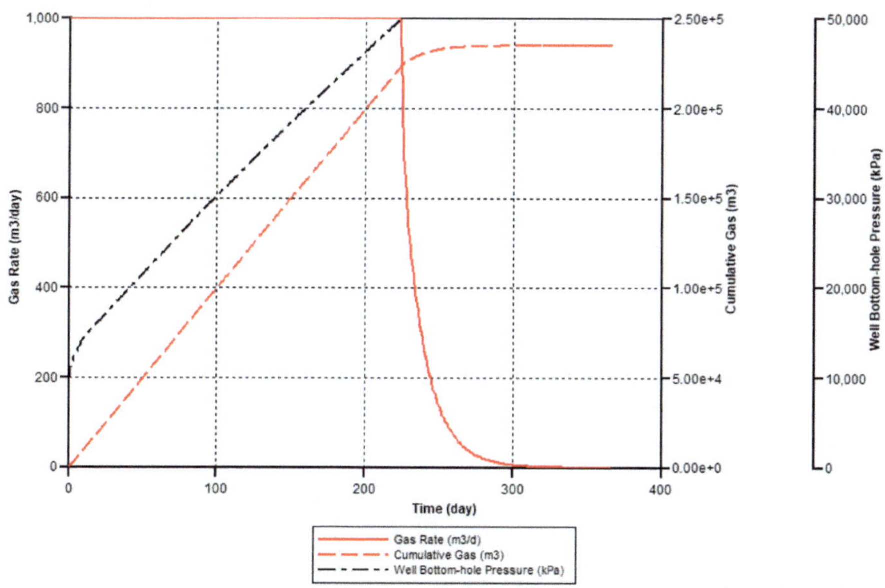

Fig. 3.10 Simulation result for borehole#4 showing gas rate, cumulative gas volume and well bottom-hole pressure

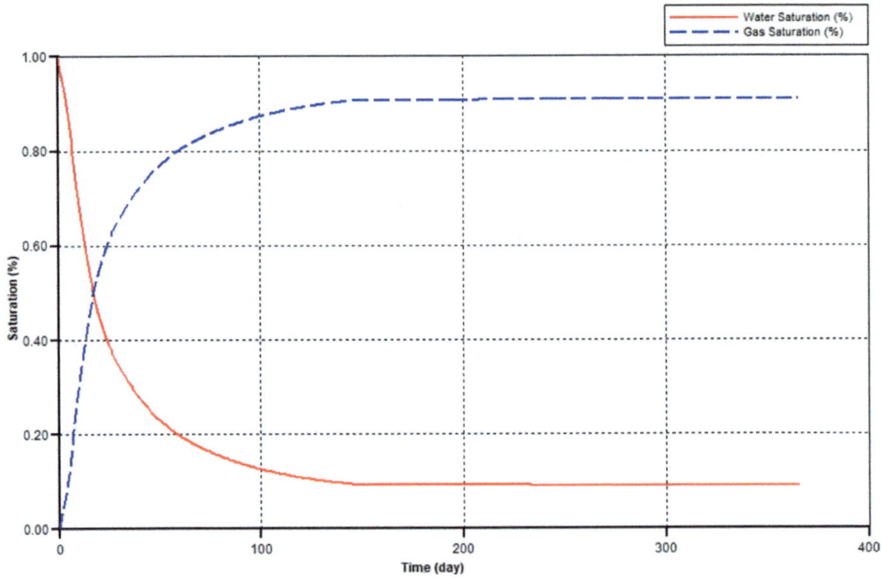

Fig. 3.11 Simulation graph of borehole#1 showing water saturation and gas saturation

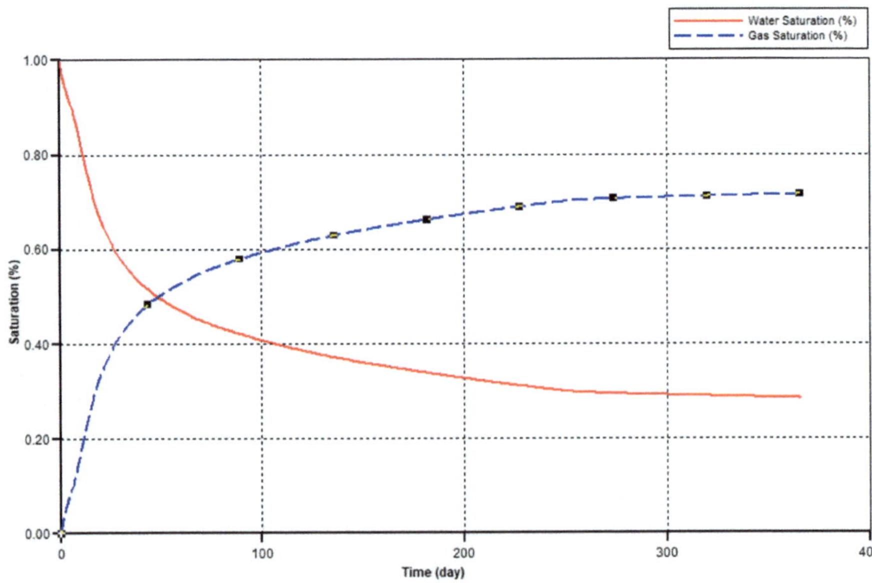

Fig. 3.12 Simulation graph of borehole#2 showing water saturation and gas saturation

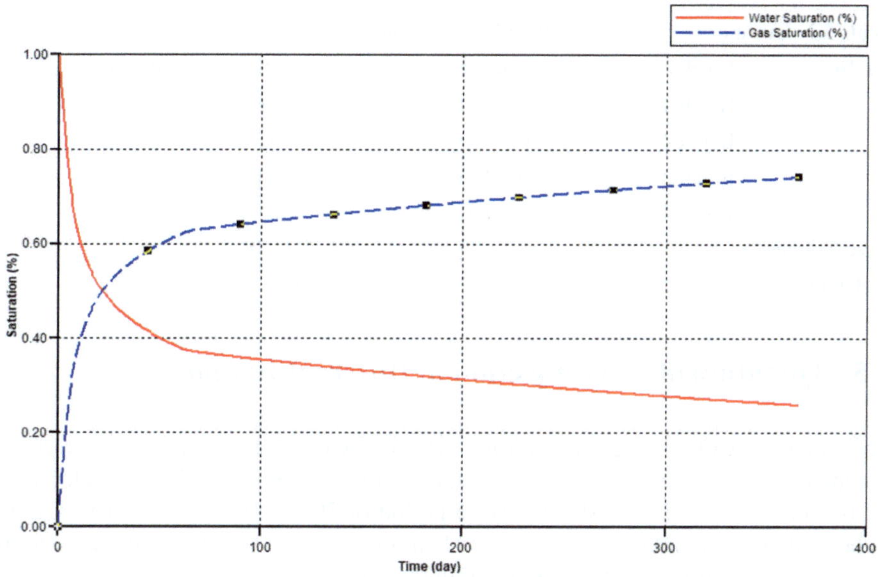

Fig. 3.13 Simulation graph of borehole#3 showing water saturation and gas saturation

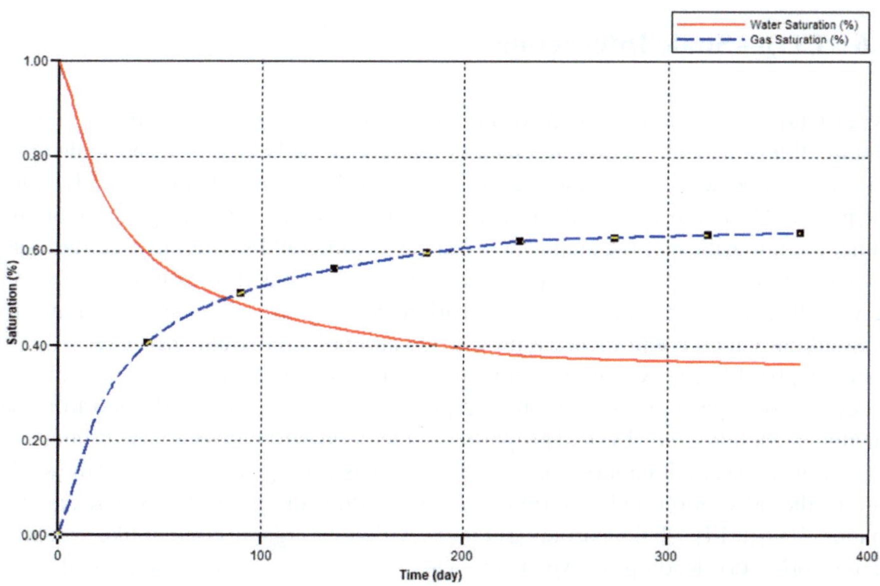

Fig. 3.14 Simulation graph of borehole#4 showing water saturation and gas saturation

Table 3.3 Quantum of CO_2 injected in different Boreholes

Si. No.	Borehole	Time (days)	Cumulative gas volume (m^3)
1	Borehole#1	300–400	1.5×10^5
2	Borehole#1	300–400	2.7×10^5
3	Borehole#1	300–400	0.61×10^5
4	Borehole#1	300–400	2.4×10^5
Total volume (m^3			7.21×10^5
Total injected CO_2 (ton)			1296.3

3.5 Quantum of CO_2 Injection in Different Boreholes

The amount of CO_2 injected, as determined by reservoir simulation results for various boreholes, is shown in Table 3.3. The total cumulative volume of CO_2 injected into all boreholes was 7.21×10^5 m^3 over a period of 300–400 days [15]. The results indicate that approximately 1296.3 tons of anthropogenic CO_2 can be sequestered in the shale reservoir of the study area within this time frame.

3.6 CO_2—Shale Interaction

When CO_2 is injected into a shale reservoir, it can interact with both the rock matrix and the fluids present, leading to changes that affect the behavior of the shale [12–15]. These interactions begin at the molecular level, where CO_2 may dissolve into the pore water and react with the minerals that make up the shale. Depending on the mineral composition of the shale, CO_2 can cause mineral dissolution, precipitation, or both, altering the chemical environment within the rock. For example, CO_2 can react with calcium carbonate or other carbonate minerals in the shale, leading to dissolution, which could increase pore space and potentially enhance CO_2 storage capacity [10, 13, 20]. At the same time, new minerals might precipitate, potentially reducing pore space or changing the permeability of the rock. In addition to these chemical reactions, the physical properties of the shale can change as CO_2 is injected. One of the most significant physical changes is the swelling or shrinkage of the shale due to the adsorption and desorption of CO_2 within the pore spaces and organic matter [10–12, 14]. Shale, particularly those with high organic content like kerogen, can absorb CO_2, leading to expansion. This swelling can impact the porosity and permeability of the shale, either improving or hindering the ability of CO_2 to migrate through the rock. On the other hand, desorption, where CO_2 is released from the shale matrix, can cause the rock to shrink, potentially creating fractures that enhance permeability. The balance between these processes is crucial in determining the overall effectiveness of CO_2 storage in shale formations [10–15].

Mechanical changes in the shale also occur due to the injection of CO_2, particularly in response to changes in pressure and fluid composition. As CO_2 is injected, the increase in pore pressure can induce stress changes in the shale, which may lead to deformation or fracturing. In some cases, this mechanical alteration can improve the storage potential by creating additional pathways for CO_2 to migrate. However, excessive fracturing or stress changes may compromise the structural integrity of the reservoir, potentially leading to leakage or reduced storage efficiency [12, 13, 15]. Additionally, the chemical weakening of the rock from CO_2-induced reactions can affect its mechanical strength, making the shale more susceptible to deformation under stress.

Together, the chemical, physical, and mechanical interactions between shale and CO_2 must be carefully considered in any CO_2 sequestration project. Understanding these processes helps optimize CO_2 storage in shale formations, ensuring that the reservoir can safely and efficiently trap CO_2 over long periods while minimizing risks associated with leakage or loss of storage capacity.

3.7 Conclusions

The experimental and simulation studies on carbon dioxide injection and gas recovery in the Chalbalpur-Mahishmura regions of the Raniganj Field, India, provide valuable insights into the potential of shale formations for CO_2 sequestration and enhanced gas recovery. The findings highlight the significance of shale properties such as porosity, permeability, mineral composition, and mechanical strength, as these characteristics significantly influence CO_2 storage and gas recovery efficiency. The interaction between CO_2 and the shale, involving chemical reactions, adsorption and desorption processes, and mechanical changes under pressure, plays a crucial role in determining reservoir performance. Simulation results showed gas saturation in boreholes ranging from 60 to 90%, indicating successful CO_2 injection. The cumulative gas volumes varied across boreholes due to differences in porosity and permeability, with higher porosity regions offering greater CO_2 storage potential. Over a span of 300–400 days, approximately 7.21×10^5 m^3 of CO_2 was sequestered, demonstrating the reservoir's capacity for long-term storage. These results emphasize the need for accurate modelling, ongoing monitoring, and effective reservoir management to optimize gas recovery while addressing risks such as structural instability or leakage. This research contributes valuable knowledge for advancing carbon capture and storage technologies, particularly in shale reservoirs.

References

1. Chalmers GRL, Ross DJK, Bustin RM (2012) Geological controls on matrix permeability of Devonian gas shales in the Horn River and Liard basins, northeastern British Columbia, Canada. Int J Coal Geol 103:120–131. https://doi.org/10.1016/j.coal.2012.05.006
2. Ghanizadeh A, Gasparik M, Amann-Hildenbrand A, Gensterblum Y, Krooss BM (2013) Lithological controls on matrix permeability of organic-rich shales: an experimental study. Energy Procedia 40:127–136. https://doi.org/10.1016/j.egypro.2013.08.016
3. Dalton JW, Virginia W (2012) Technique development to measure reservoir properties in unconventional shale core samples department of petroleum and natural gas engineering
4. Konstantinovskaya E, Rutqvist J, Malo M (2014) CO_2 storage and potential fault instability in the St. Lawrence Lowlands sedimentary basin (Quebec, Canada): insights from coupled reservoir-geomechanical modeling. Int J Greenh Gas Control 22:88–110. https://doi.org/10.1016/j.ijggc.2013.12.008
5. Wang L, Torres A, Xiang L, Fei X, Naido A, Wu W (2015) A technical review on shale gas production and unconventional reservoirs modeling 141–151
6. Rivard C, Lavoie D, Lefebvre R, Séjourné S, Lamontagne C, Duchesne M (2014) An overview of Canadian shale gas production and environmental concerns. Int J Coal Geol 126:64–76. https://doi.org/10.1016/j.coal.2013.12.004
7. Soroush H, Rasouli V, Tokhmechi B (2010) A data processing algorithm proposed for identification of breakout zones in tight formations: a case study in Barnett gas shale. J Pet Sci Eng 74(3–4):154–162. https://doi.org/10.1016/j.petrol.2010.08.012
8. Zhou J, Liang G, Deng T, Zhou S, Gong J (2017) Coalbed methane production system simulation and deliverability forecasting: coupled surface network/wellbore/reservoir calculation. Int J Chem Eng, https://doi.org/10.1155/2017/8267529
9. Liu Y, Tang D, Xu H, Li S, Tao S (2018) The impact of coal macrolithotype on hydraulic fracture initiation and propagation in coal seams. J Nat Gas Sci Eng 56:299–314. https://doi.org/10.1016/j.jngse.2018.06.013
10. Hazarika S, Boruah A (2022) Supercritical CO_2 (SCO_2) as alternative to water for shale reservoir fracturing. Mater Today: Proc 50:1754–1757
11. Hazarika S, Boruah A, Saraf S (2023) Modeling and simulation study of CO_2 fracturing technique for shale gas productivity: a case study (India). Arab J Geosci 16(7):408
12. Boruah A, Prusty BK (2023) Influence of mineralogy, organic matter and thermal maturity on the diagenetic evolution of pore systems in the shales of Cambay and Raniganj Basin, India. J Geol Soc India 99(11):1577–1585
13. Boruah A, Rasheed A, Mendhe VA, Ganapathi S (2019) Specific surface area and pore size distribution in gas shales of Raniganj Basin, India. J Petrol Explor Prod Technol 9:1041–1050
14. Boruah A, Ganapathi S (2015) Microstructure and pore system analysis of Barren Measures shale of Raniganj field, India. J Nat Gas Sci Eng. https://doi.org/10.1016/j.jngse.2015.05.042
15. Boruah A (2024) Petrophysical characterization and reservoir simulation for CO_2 sequestration in the shale reservoir. In: 85th EAGE annual conference & exhibition (including the workshop programme), vol 2024, no 1, pp 1–5. European Association of Geoscientists & Engineers. https://doi.org/10.3997/2214-4609.2024101397
16. Hou X, Zhu Y, Chen S, Wang Y (2017) Gas flow mechanisms under the effects of pore structures and permeability characteristics in source rocks of coal measures in Qinshui Basin, China. Energy Explor Exploit 35(3):338–355. https://doi.org/10.1177/0144598717700080
17. Gaurav A, Dao EK, Mohanty KK (2012) Evaluation of ultra-light-weight proppants for shale fracturing. J Pet Sci Eng 92–93:82–88. https://doi.org/10.1016/j.petrol.2012.06.010
18. Aminian K, Ph D, Borujeni AT, Virginia W (2014) Poroelastic effect on gas transport and storage in organic rich shale reservoir Bahiya Jabbar thesis submitted to the college of engineering and mineral resources at West Virginia University in partial fulfillment of the requirements for the degree of Mas

19. Zhang P, Lu S, Li J, Chen C, Xue H, Zhang J (2018) Petrophysical characterization of oil-bearing shales by low-field nuclear magnetic resonance (NMR). Mar Pet Geol 89(7):775–785. https://doi.org/10.1016/j.marpetgeo.2017.11.015
20. Guo T, Zhang S, Qu Z, Zhou T, Xiao Y, Gao J (2014) Experimental study of hydraulic fracturing for shale by stimulated reservoir volume. Fuel 128:373–380. https://doi.org/10.1016/j.fuel.2014.03.029

Chapter 4
CO_2 Fracturing as an Alternative of Hydraulic Fracturing for Shale Gas Production

Abstract Fracturing shale gas reservoirs with carbon dioxide (CO_2) represents an efficient technique in the extraction process. Enhancing the effectiveness of hydrofracturing in these reservoirs necessitates a precise analysis of their heterogeneous nature, mandating the use of 3D reservoir modelling. This modelling approach demonstrates the application of reservoir simulation techniques, crucial for forecasting successful fracturing methods in shale gas recovery. Conducting multi-stage fracturing tests on unconventional reservoirs using this model allows for the assessment of heterogeneity and variations in fracture distance, considering factors like fracture half-length, spacing, and conductivities. Employing CO_2 as a fracturing fluid, a simulation study explores diverse conditions. The findings from simulation and modelling indicate a notable enhancement in shale gas recovery rates, exhibiting an increase ranging between approximately 3–7%. These improvements are linked to alterations in fracture half-length, spacing, and the number of fractures.

Keywords CO_2 fracturing · Shale gas production · Simulation · CO_2 geosequestration

4.1 Introduction

With the increasing energy demand, the requirement of unconventional hydrocarbon resources is also increasing [1]. In the US, shale gas production has increased since 2010 from about 6.16 trillion cubic feet and exceeded 20 trillion cubic feet in 2020. It is expected to reach about 35 trillion cubic feet in 2050. Forty-nine percent of the total US gas production is expected to come from shale reservoirs by 2035 [2]. Shale gas is the in-situ generation of natural gas that exists as both adsorbed and free gas in organic-rich shales [3, 4]. For a conventional gas reservoir, shale is the source rock which generates hydrocarbon (oil and/or gas). This hydrocarbon migrates and accumulates in the reservoir pay [4]. Unlike the conventional hydrocarbon reservoirs, unconventional reservoirs act as source, reservoir, and caprock in the system, which has a complex reservoir dynamics, non-homogeneous microstructure and pore

A. Boruah, *CO_2 Geosequestration: Capturing Carbon for a Sustainable Future*, Springer Climate, https://doi.org/10.1007/978-3-031-81021-3_4

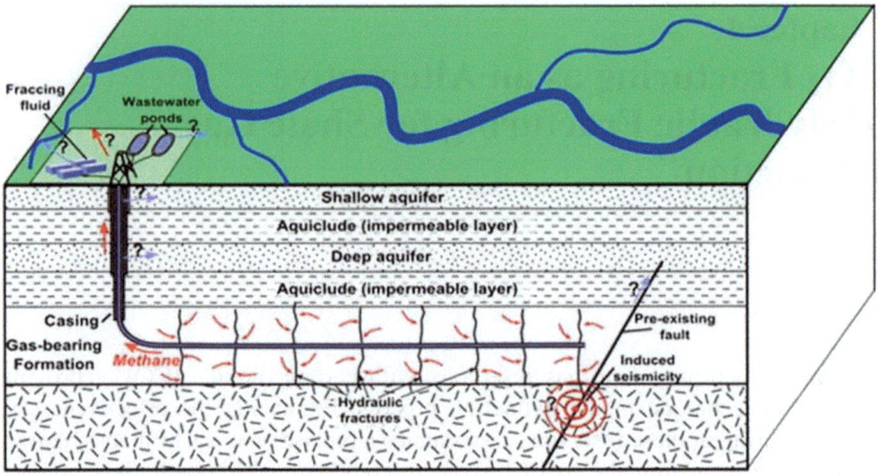

Fig. 4.1 Schematic diagram of hydraulic fracturing in shale [3] [https://en.wikipedia.org/wiki/Environmental_impact_of_fracking_in_the_United_States#/media/File:HydroFrac.png]

system [3, 4]. As the shale rocks are comprised of pores with nano to micro scale sizes and exhibit ultra-low permeability, natural gas flow is very low in shale reservoirs. Due to the heterogenous microstructure and tight nature, it requires stimulation techniques such as hydraulic fracturing for commercial shale gas exploitation [3]. But the hydraulic fracturing techniques creates micro-fractures in the shale formation by injecting water as base fluid with additives (acids, gel, etc.) and proppant (sand) under high pressure (Fig. 4.1).

Hydraulic fracturing creates artificial microfractures and enhances the permeability of the reservoir which enhances the shale gas production. However, the hydraulic fracturing is considered as non-environment friendly. It can swell due to water rock interaction. The clays are the primary component of shale, and which is a water-sensitive mineral. The use of water during artificial fracturing can create issues like formation damage, clay swelling, instability etc. [5–7]. The development of capillary end effect, water blocking, proppant placement, etc. can reduce the production rate and recovery efficiency [4]. The mineral hydration and water imbibition can cause reduce the effective permeability, while inappropriate disposal of a huge volume of flow back fluids containing chemical additives can cause environmental issues [5, 8]. Therefore, there is a demand of the environment-friendly fracturing fluids in hydrofracturing technologies. Recent researches reported uses of waterless hydrofracturing fluid to minimise the environmental issues. Laboratory-based investigations reported the liquid carbon dioxide (CO_2) as an effective hydrofracturing fluid. Recent research suggested the CO_2 fracturing as favorable technique, which has multiple benefits such as sequestering carbon, increase production, reduction of environmental hazards, etc. [6]. As relative permeability or fracturing permeability damage, formation damaged and fluid loss are critical issues in hydro- fracturing treatment, CO_2 fracturing can be used to reduce the permeability damage issues

[7, 9]. The viscosity, density, temperature, pressure, and compressibility influence the CO_2 fluid for phase behavior at the fracturing treatment [10, 11]. CO_2 interacts with shale and alters the geological, geochemical, petrophysical and geo-mechanical properties [12]. CO_2 dissolves into the formation to form carbonic acid. The shale properties alters due to acid interaction and it effects the pore structure of the shale. Comparing the water and liquid CO_2, supercritical CO_2 has a lower breakdown pressure which develops three dimensional fractures [10]. This chapter discusses the challenges of conventional hydrofracturing fluids, history of CO_2 fracturing fluid, and also the important properties of shale such are pore structure, porosity, permeability, mechanical and chemical, etc. and how they are highly influenced by interaction with CO_2 [10–13].

4.2 CO_2 Fracking and Shale Gas Recovery

Liquid CO_2 was first introduced in the petroleum industry during the early 1960s [11]. The first patent for CO_2 fracturing was issued in Canada (Canada Patent No. 1134258, U.S. Patent No. 4374545, British Patent No. 8225692), and the process was applied in the Glauconite sandstone reservoir in Canada in July 1981 [7]. In 1982, Bullen analyzed the use of liquid CO_2 as a fracturing fluid and found it improved performance in low-permeability reservoirs [7]. By 1985, Settari and colleagues developed a CO_2 fracturing model with less viscous fluids, showing through simulations that CO_2 fracturing could maintain fracture networks [12]. Luk patented a CO_2 fracturing method in 1994 that used a higher proppant concentration, and this technique was successfully used in Canada [13]. In 1993, the U.S. Department of Energy introduced waterless hydraulic stimulation methods, such as liquid CO_2, N_2 gas, and N_2 foam, at the Big Sandy gas fields in the water-sensitive Devonian shale of eastern Kentucky, to compare their gas production efficiency [7, 9–11] (Fig. 4.2).

In 2002, a field trial took place in Southern Alberta's Milk River and Medicine Hat formations to assess the economic viability of enhanced production using various CO_2-energized fracturing fluids. The results showed that these fluids notably

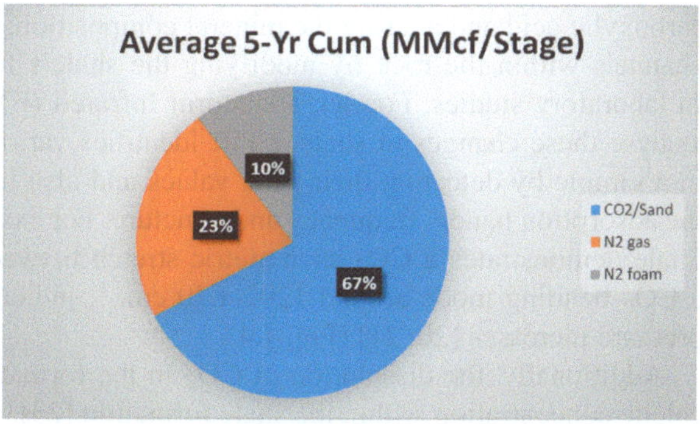

Fig. 4.2 Comparing the effect of different fracturing fluid treatment and stimulation methods on gas production from the Big Sandy reservoir [7]

increased production rates over an extended timeframe. A subsequent study in 2011, conducted by Enfrac Inc., Ferus Inc., and RPS Energy Canada Ltd., analyzed 66 horizontal wells to compare the performance of slick-water fracturing fluids with CO_2-energized alternatives. The findings indicated that CO_2-based fluids were more efficient, offering advantages such as lower water usage and quicker clean-up times [13, 14]. The successful outcomes of CO_2 liquid fracturing treatments also inspired a China-U.S. joint research initiative on CO_2-based hydraulic fracturing under the CERC-WET project [15]. Recent research on CO_2 fracturing for coalbed methane (CBM) recovery has focused on combining hydraulic slotting with liquid CO_2 injection to enhance coalbed permeability and methane recovery [16]. Simulated experiments have shown that injecting liquid CO_2 (L CO_2) into coalbeds with high gas content can significantly improve CBM recovery, with an efficiency of over 90% [17, 18].

4.3 Shale-CO_2 Interaction

4.3.1 Alteration of Pore Structure and Mineralogy

Shale is mainly composed of clay minerals, quartz, micas of tiny size, feldspar, kaolinite, carbonate, and many more, where the quartz and clay minerals are the most common minerals [19]. The pores in the shales are classified as nano to mico scale pores, organic intragranular pores, natural microfractures, and micro fractures (Fig. 4.3) [18]. Matrix pores provide the significant storage space for shale gas. These pores are found as free gas or adsorbed gas. The amount of gas present in shale reservoirs is directly proportional to the total volume of matrix pores. Fracture pores are also important for storage of free gas. This control the well productivity of the reservoir [3, 4, 11–14].

When CO_2 interacts with shale, it initially becomes trapped as free gas within the macropore structure and subsequently migrates into the nanopore system. The pore structure of shale is sensitive to pressure conditions, which in turn influence its behaviour. Upon contact, CO_2 reacts chemically with the shale, dissolving as carboxylic acid and altering the mineral composition. This reaction can create flow channels within the rock by modifying the shale's functional groups [2, 20, 21]. In laboratory studies, Fourier Transform Infrared (FT-IR) spectroscopy is used to analyse these changes in shale. FTIR identifies various mineral functional groups in a sample by detecting their peak values and also highlights vibrational shifts in the adsorption bands' frequency and structure. For example, FTIR analysis of Utica Shale demonstrates a CO_2 asymmetric stretch between 2343 and 1291 cm^{-1} and a CO_2 bending mode around 1292–670 cm^{-1}, indicating sorbed CO_2 as the CO_2 pressure increases [20, 21] (Fig. 4.4).

Additionally, the dissolution of CO_2 in the formation leads to a notable rise in calcite concentration within the shale formation [23].

Pore description	SEM image
Dissolved pore: dissolution of organic matter; corroson by alkali and acid in clay mineral. Pores are circular and elliptical	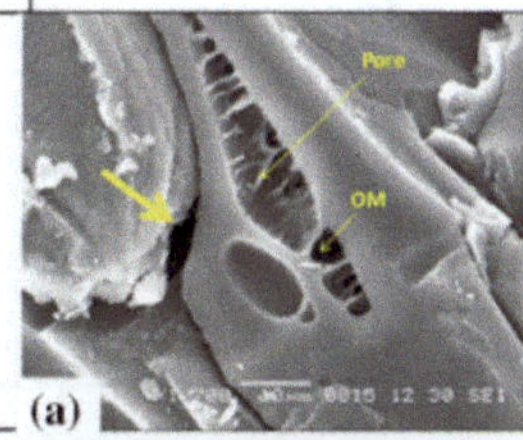
Intergranular pore: Pore at the contact of clay mineral with other mineral	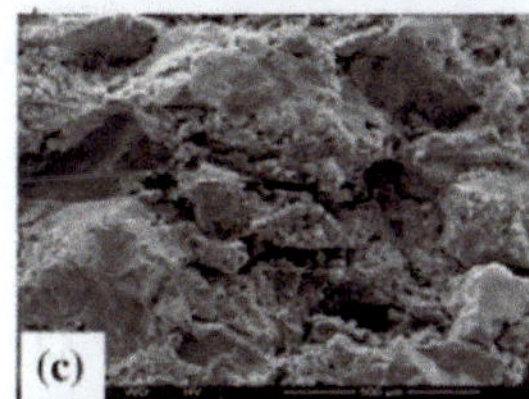
Intragranular pores: pores in chlorite. Chlorites are present in clusters, containing significant amount of pores within its structure.	
Interlayered pores: pores associated with organic matter and clay surface	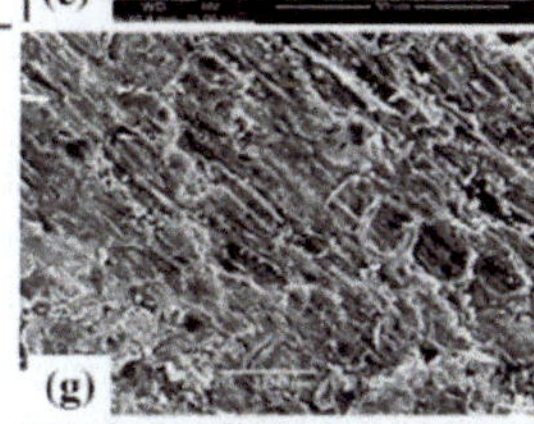
Pores in illite: appear as fibrous masses of fine crystals developing micro pores	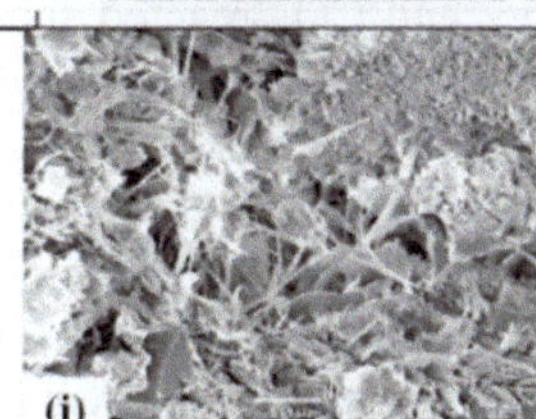
Fracture porosity: Diagenesis and tectonism can develop microcracks on the clay surface	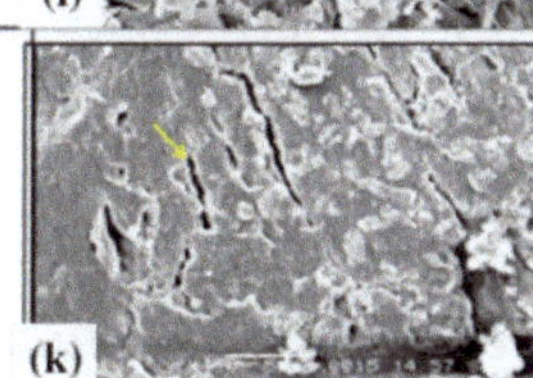

Fig. 4.3 Scanning electron microscopy images of shale pore structures [11]. http://creativecommons.org/licenses/by/4.0/

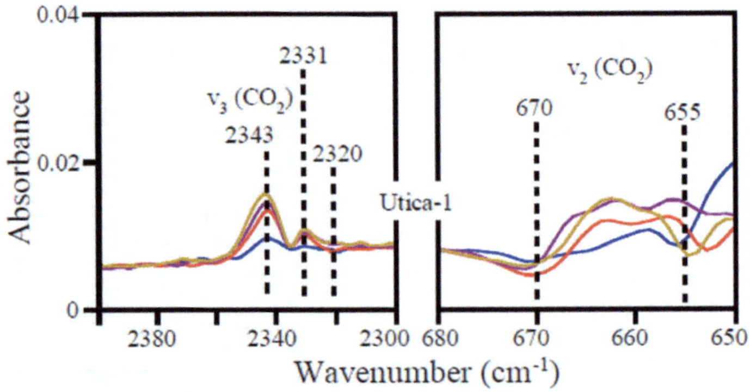

Fig. 4.4 Infrared spectrum analysis of CO_2 asymmetric stretch in the Utica Shale [22]

$$CO_2 + H_2O \leftrightarrow H_2CO_3 \leftrightarrow H+ \; +HCO_3^- \leftrightarrow 2H^+ + CO_3^{2-}$$
$$CaCO_3 + H^+HCO_3- \leftrightarrow CA_2^+ + CO_2 + H_2O$$
$$MgCO_3 + H^+HCO_3- \leftrightarrow Mg^{2+} + CO_2 + H_2O$$
$$Al_2O_3 + H^+HCO_3^- \leftrightarrow 2Al^{3+} + CO_2 + H_2O$$

The hydrolysis and carbonation processes in silicate and calcite minerals modify the microstructure and pore distribution due to diagenetic effects [23].

When chlorite and siderite are exposed to CO_2, their solubility increases, which leads to enhanced porosity and permeability. These changes are influenced by factors such as the mineral composition, duration of exposure, temperature, and pressure conditions [24]. However, CO_2 interaction with carbonate rocks tends to reduce connected porosity, resulting in a decrease in overall connected porosity in CO_2-reacted shale. Although the total porosity, pore connectivity, and surface area increase in both brine- and CO_2-reacted shales, the brine-reacted samples show no change in total porosity but experience a reduction in connected porosity [25]. On the other hand, CO_2-reacted samples display an increase in total porosity, while connected porosity decreases [8]. The low matrix permeability in these samples leads to very low flow rates and high pore-pressure gradients. Shale thickness and vertical permeability play a crucial role in determining CO_2 leakage rates, with the highest leakage occurring in shales with the greatest thickness and permeability in the nanodarcy range. This indicates that shale permeability tends to decrease after CO_2 interaction with the rock [26–32].

4.3.2 Alteration of Mechanical Properties

Shale mechanical properties, including stress, strain, stability, brittleness, etc. are essential paraments for shale gas production. Brittleness is correlated with shale mineral composition, Young's modulus, and Poisson's ratio. A rock is considered more brittle when it contains fewer clay minerals (such as kaolinite, smectite, and illite), and high quartz, feldspar, and calcite. Clay-rich rocks are more ductile and absorb more energy during fracture [3, 33]. The mechanical properties of shale refer the rock strength and stress condition which can be tested by the uniaxial compressive strength (UCS) and Young's modulus (E). When CO_2 interact with shale, then the uniaxial compressive strength (UCS) and Young's modulus (E) are reduced. The studies found that the supercritical carbon dioxide has more influence on passion ratio and young modulus as compare to supercritical CO_2 [34]. Supercritical CO_2-shale interaction may potentially enhance mechanical characteristics and brittleness of the shales. However, the shales are not homogeneous, so it is necessary to understand each shale formation on an individual basis for CO_2 fracturing.

4.3.3 Geochemical Properties Alteration

The chemical properties of shales controls the affects the shales interaction with CO_2 and its analysis by FTIR and x-ray photoelectron spectroscopy [35]. Many researchers have already analyzed the presence of oxygen-containing groups on the shale surface such as COOH, –H, and –OH bonds, which have a significant effect on CO_2 and CH_4 adsorption on carbonate minerals at high pressures when CO_2 has interacted with shale [21]. Hui et al. has shown that the chemical interactions between CO_2 and shales may reduce the oxygen-containing functional groups (C–O, O–C–O, C = O, COO–) and increased the hydrogen groups (C–C/C–H) which is the impact on total organic content in shale [35].

Water saturation in shale rock is critical for the movement of gas in a porous medium and the maintenance of fracturing fluid flow. Kerogen is hydrophobic in nature, it repels water particles, the gas trapped in organic pores [36]. The adsorption capacity of CH_4 is reduced when the moisture content rises in slit pores, while the impact of moisture on CO_2 increases. The sorption capacity of a substance is proportional to its mole fraction in a binary combination of CH_4 and CO_2. In addition, pressure drawdown removes the dispersed CH_4 molecules in the center of the kerogen slit pores, while CO_2 injection extracts the CH_4 molecules that are distributed in the middle of the kerogen slit pores. The rate of drawdown decreases as the rate of CO_2 injection rises, particularly in the initial stage of CO_2 injection [37]. The increasing moisture content reduces adsorption capacity and enhances CO_2/CH_4 adsorption selectivity [37].

4.3.4 Effect of CO_2 Injection on Pressure in Shale

In both gaseous and supercritical phases, carbon dioxide exhibits a high level of compressibility. CO_2 fracturing requires components such as storage tanks, an airtight blender, manifold trucks, and high-pressure pumps to operate effectively [38]. In cases of shallow wells, where CO_2 does not reach its supercritical phase, additional heating devices are necessary. At high pressure, CO_2 adsorption reduces the specific surface area and affects the structural and geochemical characteristics of shale during treatment [38]. The density of supercritical CO_2 increases its ability to extract organic materials from the shale, leading to higher porosity and an increased specific surface area under high-pressure CO_2 fracturing. However, over time, prolonged exposure to high pressure and temperature can extract more organic matter from the shale, potentially reducing the CO_2 storage capacity of the formation.

4.3.5 CO_2/CH_4 Competitive Adsorption in Shale

The cationic exchange and pressure affects the CO_2 adsorption and the methane adsorption is mostly determined by clay surface area [39]. CO_2 gas sorption inside the nano-pore networks of clays can cause due to cation exchange and it produces a considerable increase sorption especially at low pressure. In the absence of cationic exchange, sorption rises rapidly with increasing pressure and reaches a peak due to the dense packing of gas molecules in the pores. Strong electrostatic interactions between clay surfaces may be the cause of the significant CO_2 adsorption in clay nanopores interaction [40]. Carbon dioxide gets the supercritical phase with increasing pressure and temperature. As the supercritical CO_2 ($SCCO_2$) has a low viscosity and negligible surface tension, as well as a high adsorption capability for shale formation in comparison to methane gas (CH_4) [41]. In a specific range, the adsorption capacity of CH_4 rises with increasing pressure and decreasing temperature and moisture content [42].

4.3.6 CO_2 Injection and Fracturing

CO_2 can be injected into the shale formation with high pressure and this pressure generates fractures. In case of such fracturing no additional chemical additives are required [16]. CO_2 density, viscosity, and solubility changes with the change in temperature and pressure. CO_2 drops instantly while injecting into the reservoir. Its volume increases exponentially due to the Joule–Thomson effect [43]. At the initial stage, the CO_2 is injected into the fractures with high pressure. After that, CO_2 flow into the rock because of the pressure gradient. CO_2 enters through the rock in the third phase and it results swelling. At the final step, CO_2 pressure reduces and the gradient-driven diffusion promotes the hydrocarbon production (Fig. 4.5). The hydrocarbon is swept via cracks by bulk carbon dioxide to the production well [11, 42].

In the below mention comparing table of different fracturing fluid with respect to their reservoir conditions. All information are collect from the Scopus database has

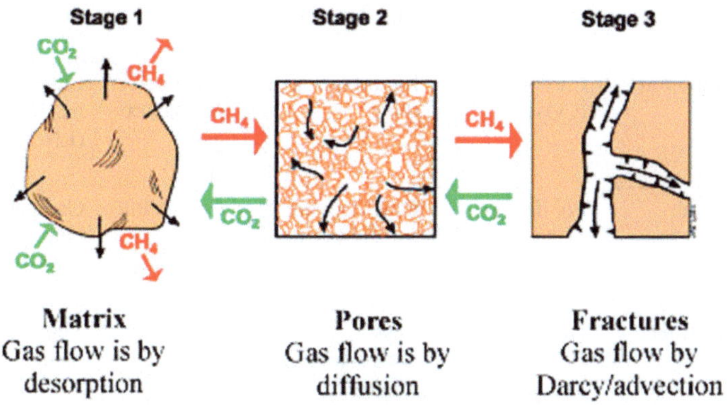

Fig. 4.5 Mechanism of CO_2 injection and fracturing in shale rock [11]

been thoroughly explored within the timeframe of January 2005 to August 2020 [11, 42] (Table 4.1).

Previous investigation on breakdown pressure and fracture surface morphology in shale with H_2O, CO_2 and N_2 fracturing treatment and analysis stated that CO_2 has the better breakdown pressure as compare with N_2 and H_2O, with H_2O having the lowest breakdown pressure [16].

4.4 Computational Fracturing Modelling Approach in Shale

A commercial simulation tool was employed to model the hydraulic fractures generated during stimulation, as well as the resulting network of natural fractures, using low-viscosity fluids. Input data, such as stress profiles and other elastic rock properties derived from geomechanical analysis, were utilized in the design process. To enhance proppant distribution, a combination of low-viscosity linear gel and slick water was used in the treatment. The linear gel was optimized for high reservoir temperatures by adjusting the concentration of ingredients sourced locally and from available water supplies. Due to the fluid's low viscosity and high closure pressures, small-mesh, high-strength proppants were selected for the design. The shale gas reservoir model was composed of x * y * z cells, representing multiple sedimentary layers within the shale formation. The simulation of fractured shale gas better aligns with real-world conditions. In Fig. 4.6, the models are broadly categorized into three types: discrete fracture models, continuous models, and hybrid models that combine both discrete and continuous fracture models.

Langmuir isothermal equation was proposed by French chemist in 1918 to analyze gas adsorption onto solid surfaces based on the molecular dynamics. This equation follows following assumptions: (I) the surface of an adsorbent is uniform while gas

Table 4.1 A comparison of different fracturing fluid with respect to their reservoir conditions [11, 42, 44, 45]

Reservoir condition	Carbon dioxide (CO_2)	Liquid nitrogen (N_2)	Liquefied petroleum gas (LPG)	Liquefied natural gas (LNG)
Formation depth	For deeper depth	Only compatible for shallow depth	Compatible for both shallow and deeper depth	Compatible for both shallow and deeper depth
Environment-Friendly	Due to loss of CO_2 into atmosphere increase the environmental pollution	Less environmental impact	Need very less water thus eliminates pollution	Very nature-friendly
Viscosity & proppant carrying capacity	Low viscosity hence poor proppant carrying capacity	Moderate viscosity and has better proppant carrying ability than CO_2	Moderate viscosity hence good proppant carrying ability	Low viscosity hence poor proppant carrying capacity
Separation process & flowback	Difficult separation process but rapid and complete flowback	Easier treatment and quick clean-up process	Simple separation process and high flowback rate	Reusable and high flowback rate
Initial cost	High costing is required for the initial establishment	Medium category initial cost is required	High initial cost	Low initial cost
Economic viability	High NPV over a long period	High NPV but less than CO_2	Lowest NPV with a short period	Moderate NPV over a short period

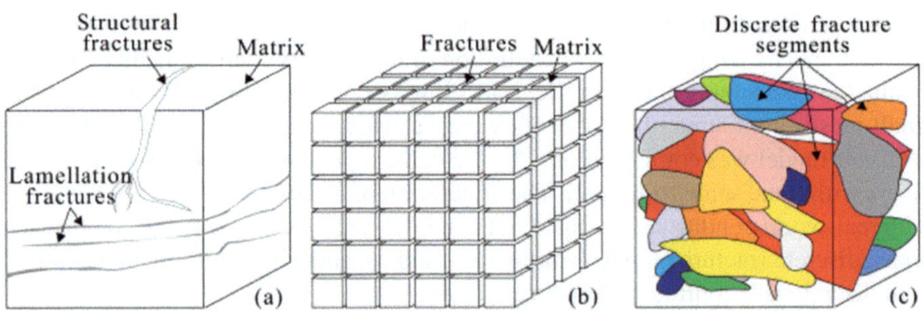

Fig. 4.6 A generalized fracture-matrix models of shale. **a** Diagram of a shale gas reservoir; **b** Dual-medium model; **c** Discrete fracture network model [17]

adsorbs to the surfaces by monolayer adsorption; (II) the process is dynamic; (III) at equilibrium, the adsorption rate is equal to the desorption rate; (IV) the rate of gas molecule adsorption on the solid surfaces increases with increasing gas pressure. The adsorption rate on the unit area of a rock surface (J_{ads}) is:

$$J_{ads} = K_a P(1 - \Theta) \tag{4.1}$$

And the desorption rate on the unit area of the rock surface (J_{des}) is:

$$J_{des} = k_d \theta \tag{4.2}$$

where: J_{ads} = gas adsorption amount on the unit area of rock surface (m/s); J_{des} = gas desorption amount on the unit area of rock surface (m/s); θ = gas coverage of the porous rock surface (dimensionless); p = gas partial pressure (Pa); k_a = gas equilibrium adsorption rate [m/(Pa s)]; k_d = gas equilibrium desorption rate (m/s).

Solving both this equation and the common Langmuir equation format is:

$$\theta : \frac{G}{G_m} = \frac{bp}{1 + bp}$$

where, G = gas adsorption amount at equilibrium (m^3/m^3); G_m = limit gas adsorption amount when gas adsorption on unit surface area is saturated (m^3/m^3); B = Langmuir adsorption equilibrium constant (1/Pa) [3]. In the Langmuir model, temperature is considered constant throughout the entire adsorption process. It is also assumed that shale gas desorption occurs instantaneously, with all desorbed gas converting to free gas. Several petroleum experts have traditionally employed the unsteady-state flow model to analyse shale gas reservoirs. Various researchers have factored in different methods and variables for specific analyses. Building on these prior research findings, the following integrated microscopic flow models are introduced and applied to examine the macroscopic flow behaviour in fractured reservoirs.

Using the continuity equation for demonstrating the gas flow in shale fractures is:

$$\frac{1}{r}\frac{\partial}{\partial r}\left(\frac{k_f}{\mu_g}\rho_g r \frac{\partial p_f}{\partial r}\right) = \frac{\partial(\phi_f \rho_g)}{\partial t} + q_{des}$$

where: P_f = pressure of micro fracture (Pa); ρ_g = gas density (kg/m^3); ϕ_f = porosity of micro fracture (dimensionless); k_f = permeability of micro fracture (m^2); q_{des} = gas mass by steady state desorption from unit reservoir volume [kg/(m^3 s)]. According to the Langmuir equation it becomes:

$$q_{des} = \rho_{gsc}(1 - \phi_f) G_L \frac{p_L}{(p_L + p_f)^2} \frac{\partial p_f}{\partial t}$$

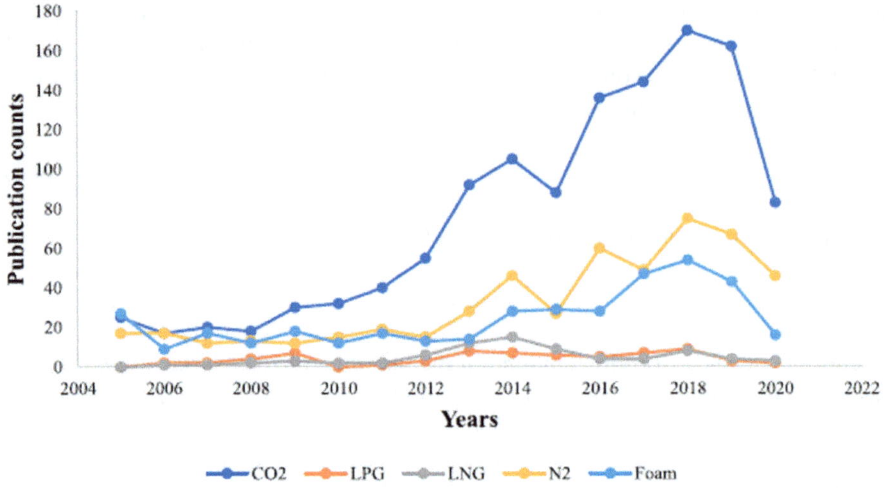

Fig. 4.7 Comparison of waterless fracturing growth from 2004 to 2022 [12, 13, 42]

where ρ_{gsc} is the gas density at standard condition (kg/m^3). This model is widely used in case of shale gas reservoirs. Figure 4.7 shows a comparison of fracturing without water from 2004 to 2022, which suggests increasing demand for CO_2 fracturing solutions.

4.5 Conclusion

CO_2 fracturing instead of using water and permanent storage of CO_2 can be useful for CO_2 mitigation from the atmosphere and it utilization for the profit of energy exploration. Supercritical CO_2 (SCO_2) is an environmentally friendly. The SCO_2 injection facilitates in enhancing the production rate and also increases the recovery of the residual gas by the altering the shale properties.

References

1. Wennersten R, Sun Q, Li H (2015) The future potential for carbon capture and storage in climate change mitigation—an overview from perspectives of technology, economy and risk. J Clean Prod 103:724–736
2. Szulczewski ML, MacMinn CW, Herzog HJ, Juanes R (2012) Lifetime of carbon capture and storage as a climate-change mitigation technology. Proc Natl Acad Sci 109(14):5185–5189
3. Environmental impact of fracking in the United States (2024) In Wikipedia. https://en.wikipedia.org/wiki/Environmental_impact_of_fracking_in_the_United_States
4. Boruah A, Ganapathi S (2015) Microstructure and pore system analysis of barren measures shale of Raniganj field, India. J Nat Gas Sci Eng 26:427–437

5. Zhang D, Song J (2014) Mechanisms for geological carbon sequestration. Procedia IUTAm 10:319–327
6. Huppert HE, Neufeld JA (2014) The fluid mechanics of carbon dioxide sequestration. Annu Rev Fluid Mech 46(1):255–272
7. Gunter WD, Bachu S, Benson S (2004) The role of hydrogeological and geochemical trapping in sedimentary basins for secure geological storage of carbon dioxide. Geol Soc London, Spec Publ 233(1):129–145
8. Aminu MD, Nabavi SA, Rochelle CA, Manovic V (2017) A review of developments in carbon dioxide storage. Appl Energy 208:1389–1419
9. Ramadhan R, Tapanya C, Akamine T, Leelasukseree C, Tangparitkul S (2024) CO_2 trapping dynamics in tight sandstone: insights into trapping mechanisms in Mae Moh's reservoir. J Environ Manage 370:122442
10. Mitchell AC, Dideriksen K, Spangler LH, Cunningham AB, Gerlach R (2010) Microbially enhanced carbon capture and storage by mineral-trapping and solubility-trapping. Environ Sci Technol 44(13):5270–5276
11. Boruah A, Rasheed A, Mendhe VA, Ganapathi S (2019) Specific surface area and pore size distribution in gas shales of Raniganj Basin, India. J Pet Explor Prod Technol 9:1041–1050
12. Hazarika S, Boruah A (2022) Supercritical CO_2 (SCO_2) as alternative to water for shale reservoir fracturing. Mater Today: Proc 50:1754–1757
13. Hazarika S, Boruah A, Saraf S (2023) Modeling and simulation study of CO_2 fracturing technique for shale gas productivity: a case study (India). Arab J Geosci 16(7):408
14. Iddphonce R, Wang J, Zhao L (2020) Review of CO_2 injection techniques for enhanced shale gas recovery: prospect and challenges. J Nat Gas Sci Eng 77:103240
15. Boruah A, Prusty BK (2023) Influence of mineralogy, organic matter and thermal maturity on the diagenetic evolution of pore systems in the shales of Cambay and Raniganj Basin, India. J Geol Soc India 99(11):1577–1585
16. Hazarika S, Boruah A, Kumar H (2023) Study of pore structure of shale formation for CO_2 storage. Mater Today: Proc.https://doi.org/10.1016/j.matpr.2023.06.014
17. Singh S, Boruah A, Devaraju J (2023) CO_2 sequestration and CH_4 extraction from unmineable coal seams in Singrauli coalfield. Int J Coal Prep Utilization 1–18
18. Rasmusson M (2018) Residual and solubility trapping during geological CO_2 storage: numerical and experimental studies. Doctoral dissertation, Acta Universitatis Upsaliensis
19. Suekane T, Nobuso T, Hirai S, Kiyota M (2008) Geological storage of carbon dioxide by residual gas and solubility trapping. Int J Greenhouse Gas Control 2(1):58–64
20. Soleimani M (2017) Naturally fractured hydrocarbon reservoir simulation by elastic fracture modeling. Pet Sci 14:286–301. https://doi.org/10.1007/s12182-017-0162-5
21. Khandoozi S, Hazlett R, Fustic M (2023) A critical review of CO_2 mineral trapping in sedimentary reservoirs–from theory to application: pertinent parameters, acceleration methods and evaluation workflow. Earth-Sci Rev 104515
22. Bachu S, Gunter WD, Perkins EH (1994) Aquifer disposal of CO_2: hydrodynamic and mineral trapping. Energy Convers Manage 35(4):269–279
23. Xu T, Apps JA, Pruess K (2004) Numerical simulation of CO_2 disposal by mineral trapping in deep aquifers. Appl Geochem 19(6):917–936
24. Bachu S (2000) Sequestration of CO_2 in geological media: criteria and approach for site selection in response to climate change. Energy Convers Manage 41(9):953–970
25. Nassabeh M, You Z, Keshavarz A, Iglauer S (2024) Sub-surface geospatial intelligence in carbon capture, utilization and storage: a machine learning approach for offshore storage site selection. Energy 305:132086
26. Luboń K (2021) Influence of injection well location on CO_2 geological storage efficiency. Energies 14(24):8604
27. Solomon S (2007) Carbon dioxide storage: geological security and environmental issues—case study on the sleipner gas field in Norway. Bellona report 128
28. Ramírez A, Hagedoorn S, Kramers L, Wildenborg T, Hendriks C (2010) Screening CO_2 storage options in the Netherlands. Int J Greenhouse Gas Control 4(2):367–380

29. Welkenhuysen K, Ramírez A, Swennen R, Piessens K (2013) Strategy for ranking potential CO_2 storage reservoirs: a case study for Belgium. Int J Greenhouse Gas Control 17:431–449
30. Raza A, Rezaee R, Gholami R, Bing CH, Nagarajan R, Hamid MA (2016) A screening criterion for selection of suitable CO_2 storage sites. J Nat Gas Sci Eng 28:317–327
31. Sing KSW (1982) Reporting physisorption data for gas/solid systems. Pure Appl Chem 54(11):2201–2218. https://doi.org/10.1351/pac198254112201
32. Sendula E et al (2017) Experimental study of CO_2-saturated water—Illite/Kaolinite/Montmorillonite system at 70–80 °C, 100–105 bar. Energy Procedia 114(November 2016):4934–4947. https://doi.org/10.1016/j.egypro.2017.03.1635
33. Mouzakis KM et al (2016) Experimental study of porosity changes in shale caprocks exposed to CO_2-saturated brines I: evolution of mineralogy, pore connectivity, pore size distribution, and surface area. Environ Eng Sci 33(10):725–735. https://doi.org/10.1089/ees.2015.0588
34. Tarkowski R, Wdowin M (2011) Études pétrographique et minéralogique de l'impact de l'injection de CO_2 dans les roches réservoirs de la basse plaine de Pologne. Oil Gas Sci Technol 66(1):137–150. https://doi.org/10.2516/ogst/2011005
35. Miller QRS et al (2016) Experimental study of porosity changes in shale caprocks exposed to carbon dioxide-saturated brine II: insights from aqueous geochemistry. Environ Eng Sci 33(10):736–744. https://doi.org/10.1089/ees.2015.0592
36. Zhang B, Shan B, Zhao Y, Zhang L (2020) Review of formation and gas characteristics in shale gas reservoirs. Energies 13(20). https://doi.org/10.3390/en13205427
37. Fatah A, Bennour Z, Ben Mahmud H, Gholami R, Hossain MM (2020) A review on the influence of CO_2/shale interaction on shale properties: implications of CCS in shales. Energies 13(12). https://doi.org/10.3390/en13123200
38. Zhang L (2019) Shale gas reservoir characteristics and microscopic flow mechanisms. In: Developments in petroleum science, vol 66. Elsevier B.V., pp 1–47
39. LaFollette RF, Hurt RS (2016) Hydraulic fracturing. In: Fossil fuels: current status and future directions, pp 229–288. https://doi.org/10.1142/9789814699983_0009
40. Barati R, Liang JT (2014) A review of fracturing fluid systems used for hydraulic fracturing of oil and gas wells. J Appl Polym Sci 131(16):1–11. https://doi.org/10.1002/app.40735
41. Howarth RW, Santoro R, Ingraffea A (2011) Methane and the greenhouse-gas footprint of natural gas from shale formations. Clim Change 106(4):679–690. https://doi.org/10.1007/s10584-011-0061-5
42. Song X et al (2019) Fracturing with carbon dioxide: from microscopic mechanism to reservoir application. Joule 3(8):1913–1926. https://doi.org/10.1016/j.joule.2019.05.004
43. Lancaster G, Lou Sinal M (1986) Liquid CO_2 fracturing advantages and limitations. Prepr Annu Tech Meet Petrol Soc CIM 2:425–435. https://doi.org/10.2118/87-05-01
44. Qian Y, Guo P, Wang Y, Zhao Y, Lin H, Liu Y (2020) Advances in laboratory-scale hydraulic fracturing experiments. Adv Civil Eng 2020(1). https://doi.org/10.1155/2020/1386581
45. Wang J, Elsworth D, Wu Y, Liu J, Zhu W, Liu Y (2018) The influence of fracturing fluids on fracturing processes: a comparison between water, oil, and SC-CO_2. Rock Mech Rock Eng 51(1):299–313. https://doi.org/10.1007/s00603-017-1326-

Chapter 5
CO_2 Sequestration and CH_4 Extraction from Unmineable Coal Seams

Abstract This chapter explains CO_2 sequestration and CH_4 extraction potential of unmineable coal seams. The coal seams of Barakar formation of Singrauli coalfield in India was considered for this study. This sequestration process has dual benefits offers a promising approach to mitigating greenhouse gas emissions in the meantime improving natural gas recovery through coal bed methane (CBM) exploration. The studied coal seams show significant storage capacities, and estimated maximum sorption values are 10,758 Mm^3 for CO_2 and 3215 Mm^3 for CH_4. The results highlight the feasibility of utilizing unmineable coal seams for carbon sequestration and methane extraction which can contribute to both climate change mitigation and energy development.

Keywords CO_2 sequestration · CH_4 extraction · Singrauli coalfield · Unmineable coal seams · Sorption · CO_2 storage capacities

5.1 Introduction

Coal is the most abundant fossil fuel in India and contributing almost 56% of the country's energy need [1]. A total 361.41 Bt of coal resource have been estimated in India as on 01.04.2022 up to maximum depth of 1200 m. Almost 9% of coal are present in the depth of 600–1200 m [1, 2]. By 2040, global primary energy demand will increase from 6% today to 11%. It is projected that almost 42% of India's energy demand will be meet by coal. This increase in fossil fuel consumption and demand leads to CO_2 emissions up to 5 Gt by 2040. It is also expected that India's share of global carbon emissions will increase from 7% now to 14% by 2040 [1–3]. The Directorate General of Hydrocarbons (DGH) of India estimates that coal deposits in India (in twelve states of India covering an area of 35,400 km^2) contain approximately 2600 BCM of CBM (DGH India, 2019). Coal in the Indian basins are high volatile to low-volatile bituminous and high in ash content (10–40%) and gas content, 3–16 m^3/t depending on the rank of the coal, depth of burial, and geotectonic settings of the basins. The carbon dioxide (CO_2) can be injected to recovery of coal bed methane

A. Boruah, *CO_2 Geosequestration: Capturing Carbon for a Sustainable Future*, Springer Climate, https://doi.org/10.1007/978-3-031-81021-3_5

from the coal seams with favourable parameters at supercritical conditions. The Carbon dioxide displaces the methane from coal bed because of its four-time higher binding capacity with coal than methane [4, 5]. Carbon dioxide (CO_2) adsorption and desorption process of methane (CH_4) displaces methane (CH_4) and the adsorbed CO_2 permanently to store in coal bed. This process is called Enhanced Coal Bed Methane (ECBM) recovery. Two molecules of CO_2 can be sequestered for every released molecule of methane [5]. Deep and unmineable Coal seam can be used effectively for methane extraction and a geological sink for carbon dioxide [1].

Coal quality parameters in terms of rank, grade, composition, mineralogy, ash content, thermal maturity, etc. have significant role in determining its suitability for CO_2 sequestration and CH_4 sorption in the matrix of coal. The petrophysical parameters like porosity and permeability of coal affects the storage capacity and flow mechanism of gases. Coals with high porosity and high permeability (interconnected pore spaces) can store larger volumes of CO_2 and CH_4. The gas flow in the coal layer is highly affected by permeability, high gas flow movement of CO_2 within the coal matrix can be expected in high permeable coals. The adsorption mechanism of CO_2 onto the coal surface also helps in the sequestration potential. Its adsorption capacity is influenced by coal rank, maceral composition, and surface area. The coals with higher rank generally have greater CO_2 and CH_4 sorption capacities due to their molecular structure. The mineral composition of coal also contributes its ability to sequester CO_2 and absorb CH_4. For example, minerals such as clays and carbonate react with CO_2 and form stable compounds to improve long-term storage. Some diagenesis effects may reduce porosity and permeability, limiting gas storage. The mineral composition of must be assessed to understand its suitability for sequestration. The carbon content in coal is very high and the higher carbon content implies a greater capacity for CO_2 storage. However, the structural integrity of coal is important in ensuring that it can maintain its stability over time without significant physical changes which can affect its storage potetial. Therefore, appropriate coal property selection is important for CO_2 sequestration which involves evaluating porosity, permeability, adsorption capacity, mineral composition, carbon content, and structural integrity. This ensures optimized storage potential and long-term stability of CO_2 storage sites. Mares et al. (2009) stated that low-rank coal has a greater ability to adsorb CO_2 compared to high-rank coal. The subbituminous coal shows a CO_2 to CH_4 volumetric ratio as high as 7:1 and high-rank coals typically exhibit a 2:1 ratio. However, most research studied about high-rank coal and there is limited study available on low-rank coal [1–7]. The present study aims to investigate the CO_2 storage potential in low-rank coal by examining samples from eight coal seams in the Moher Main Basin, located in Madhya Pradesh, India.

Methane is formed during the process of coalification, and methane generation is depend on pressure and geothermal conditions. It is reported that anthracite coal can generate up to 30,000 Scf/ton of methane [5]. The generated methane either remain trapped within the coal layers or escape to surrounding rocks. When methane is trapped within the coal bed, it is referred to as coal bed methane (CBM). The retention of methane depends on various parameters such as coal matrix properties, geothermal temperature, and pressure [5]. Coal matrix properties are affected by the rank of coal, ash content, moisture content, maceral composition, cleats, porosity, and permeability [5]. Macerals are small organic particles that aid in cleat formation within the coal [5]. Unlike the conventional gas reservoirs, where source and reservoir rocks differ, in CBM, coal serves as both the source and reservoir for methane. Methane retention in CBM differs from conventional reservoirs, where gas can be stored in a compressed, free state. In coal formations, the methane is stored primarily in an adsorbed state [5]. Several factors, including coal rank, porosity, permeability, and maceral content, control CBM and CO_2 sequestration potential [5, 6]. Previous researchers [5] have investigated the desorption characteristics of CBM reservoirs and their controlling factors.

5.2 Important Parameters for CO_2 Storage

5.2.1 Porosity

Absolute porosity (ϕ) refers to the ratio between the total void volume (V_0) and the bulk volume (V_b) of a sample matrix.

$$\Phi = V_0/V_b\Phi = V_0/V_b\Phi = V_0/V_b$$

During the coalification process, methane can accumulate within the available spaces in the coal matrix. Coal features a dual pore system consisting of macropores and micropores. Macropores are formed by cleats and other natural fractures, where free methane and water are stored [7–9]. These macropores provide pathways for water and methane transport within the coal matrix, but they play a minor role in methane storage. Less than 10% of the methane present in coal seams is found in macropores. On the other hand, micropores are extremely small capillaries and cavities of molecular size. Methane in these micropores is stored in an adsorbed state, so tightly bound that it behaves like a liquid. Estimates suggest that 98% of the total methane is adsorbed within micropores [7]. While coal's porosity may be as low as 2% in macropores, the storage capacity in micropores can be equivalent to 20% porosity in sandstone, assuming 100% gas saturation under similar geological conditions. Micropores in coal are approximately 40 Å in diameter, and methane molecules in the micropores are held by Van der Waals forces in the adsorbed phase. The reverse of this process is called desorption [8–10].

5.2.2 Permeability

Permeability is a crucial factor in evaluating the economic viability of a reservoir, as it measures a rock's ability to transport fluids through a network of interconnected pores [11–15]. Accurate permeability measurements using core samples are challenging because permeability in coal depends on the sample size [13]. Since factors affecting porosity also influence permeability, permeability values are often correlated with porosity and extrapolated across reservoirs. During coalbed methane (CBM) production, the permeability of the coal matrix changes due to three mechanisms: the Klinkenberg effect, matrix shrinkage, and effective stress. The Klinkenberg effect and matrix shrinkage increase permeability, while effective stress reduces it. Several methods have been developed to analyze and evaluate gas adsorption in coal matrices, and some of the key theoretical and analytical approaches are summarized below [14, 15].

5.3 Resource Estimation for CO_2 Storage

Estimating CO_2 storage in coal seams involves assessing the capacity of unmineable or economically unviable coal seams to adsorb and store carbon dioxide. Coal seams, particularly those that are unmineable due to factors like depth, thin seams, or high stripping ratios, offer potential for CO_2 sequestration through enhanced coalbed methane recovery (ECBM) or direct CO_2 storage [2, 5]. The process of resource estimation involves several key steps:

5.3.1 Coal Seam Characteristics Assessment

The thickness, depth, and areal extent of the coal seams are critical. Accurate mapping through borehole data, seam contours, and geological models is essential for identifying viable coal seams for CO_2 storage. Coal type and rank influence the gas adsorption capacity. Coals with high vitrinite content generally have greater gas storage potential). The capacity for CO_2 storage depends on the coal's porosity and permeability. Low permeability can limit gas movement, but adsorbed CO_2 in micropores remains securely stored.

5.3.2 Sorption Capacity Estimation

The Langmuir isotherm is used to model the adsorption behavior of gases in coal seams. It provides two key parameters: Langmuir Pressure Constant (P_L): This defines the pressure at which half of the maximum adsorption capacity is reached. Langmuir Volume Constant (V_L): This represents the maximum volume of CO_2 that the coal seam can adsorb at a given pressure. Laboratory experiments can be performed to establish [5].

The maximum possible adsorption capacity of a coal seam can be estimated using the Langmuir equation:

$$V_{ads} = V_L \times P/P_L + P$$

where:

- V_{ads}—is the volume of CO_2 adsorbed.
- V_L—is the Langmuir volume constant.
- P—is the reservoir pressure.
- P_L—is the Langmuir pressure constant.

This provides an estimate of the maximum amount of CO_2 that can be stored in the coal seam under ideal conditions. However, the actual capacity may be lower due to factors like matrix swelling and the saturation of the coal with methane or other gases.

The storage capacity of a coal seam is affects by its pressure and temperature, as these factors determine the phase behaviour of CO_2 and its adsorption onto the coal. CO_2 injection into the seam can cause the coal matrix to swell, potentially decreasing pore space and permeability, which must be considered in capacity estimations. Additionally, the existing gas saturation within the seam (e.g., methane, nitrogen) plays a role in its capacity to adsorb CO_2. The laboratory sorption studies help in understanding the theoretical maximum adsorption values, field data from CO_2 injection operations offer more practical insights. Numerical simulation models can be used to predict the actual CO_2 storage capacity, considering:

- Geologic heterogeneity.
- Gas diffusion rates.
- CO_2 injection rates.
- Long-term storage behavior.

Popular simulation software like COMET3, TOUGH2, or GEM can model these dynamic processes and offer predictions on storage capacity under various operational conditions [5]. CO_2 Storage Potential Estimation Formula: Once key parameters such as seam thickness, areal extent, and adsorption capacity have been determined, the total CO_2 storage potential of a coal seam can be estimated as follows:

$$CO_2\,\text{Storage}\left(m^3\right) = A \times h \times \phi \times V_L \times F$$

where:

- A = Areal extent of the coal seam (m^2).
- h = Thickness of the coal seam (m).

- ϕ = Porosity.
- V_L = Langmuir volume constant.
- F = Factor for in-situ conditions (adjusts for saturation, pressure, and temperature).

5.4 Case Study: Moher Basin, India

5.4.1 Coal Quality Assessment

The research area is part of the Singrauli coal field, located in the Moher Main basin of Singrauli district, Madhya Pradesh, India. The selected coal-bearing region for the study covers approximately 30 km^2 within the larger basin, which spans 1900 km^2. The coal deposits in this basin primarily belong to the Barakar and Raniganj formations from the Permian period. Certain coal seams from the Barakar formation in the Singrauli coal fields are considered unmineable due to their greater depth and thin seams [2, 5].

In the 30 km^2 section of the study area in the Singrauli Main basin, a sorption study was conducted on methane (CH_4) and carbon dioxide (CO_2) in coal samples using a specially designed laboratory apparatus. The experimental studies were carried out on adsorption and the Langmuir isotherm for different coal samples collected from the Singrauli coalfield. Coal quality parameter, including proximate analysis and vitrinite reflectance measurements was evaluated through various laboratory tests. These tests helped in determining the key coal properties, such as moisture content (M%), ash content (A%), volatile matter (VM%), fixed carbon (FC%), and vitrinite reflectance (Ro, max), which are important for assessing its suitability for energy production and sequestration processes. In the Singrauli Coal Field's Main Basin, a thorough coal quality assessment was performed. The proximate analysis of coal samples showed moisture content ranging from 3.30 to 10.70%, volatile matter between 3.30 and 10.70%, fixed carbon from 17.69 to 59.86%, and ash content between 11.40 and 53.00%. The average values were 6.29% for moisture, 27.67% for volatile matter, 42.63% for fixed carbon, and 23.41% for ash [5]. The vitrinite reflectance of the coal samples ranges from 0.69 to 1.97, with an average of 1.20, indicating that the coal in the study area ranges from high-volatile A bituminous to low-volatile bituminous [5, 12, 13, 15] (Fig. 5.1).

The Barakar Formation coal seams in Singrauli coalfield comprises a 30 km^2 area. The estimated maximum sorption capacities is 39.50 m^3/t for CO_2 and 11.15 m^3/t for CH_4 for the study area. This capacity is equal to total capacities of 10,758.00 Mm^3 for CO_2 and 3215.00 Mm^3 for CH_4. These results indicates the high potential for CO_2 sequestration and CBM extraction in the unmineable coal seams in the Singrauli coalfield. Among the eight coal seams, the seam VI has the highest methane adsorption capacity at 616.9 Mm^3, Seam IV shows the highest carbon dioxide adsorption capacity at 2428.5 Mm^3. CO_2 injection is feasible at depths greater than 500 m, with the maximum sorption potential of the basin's current geophysical conditions

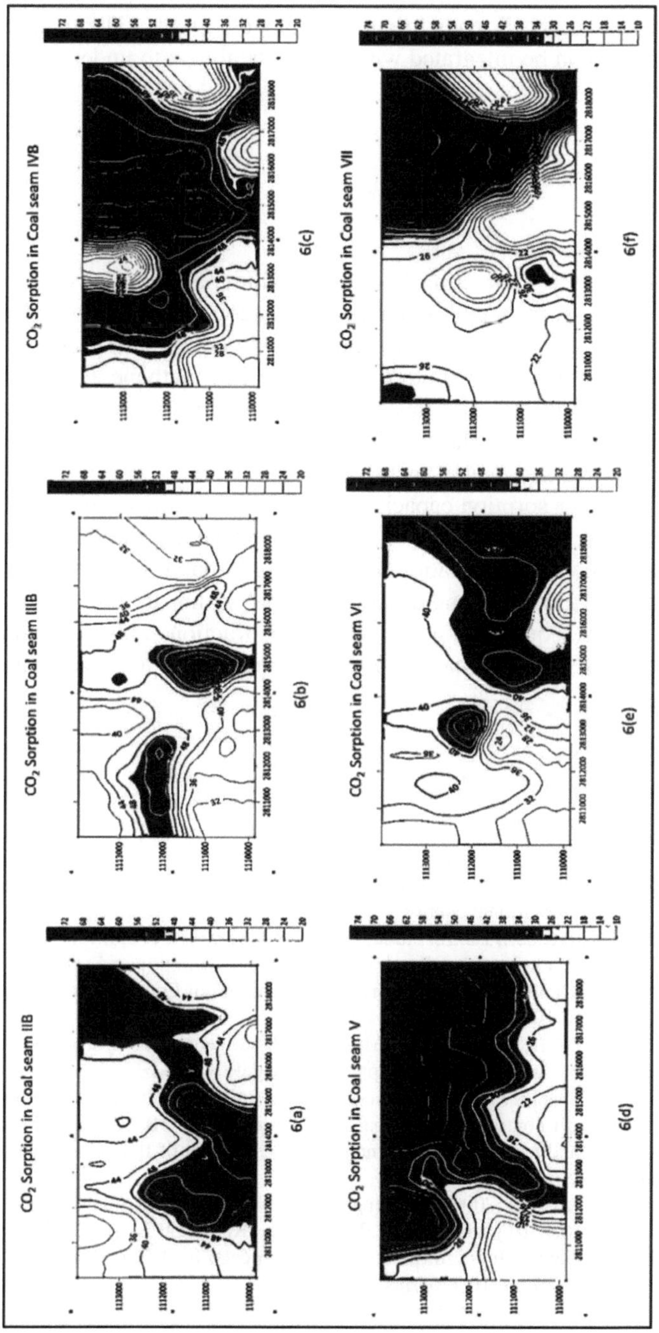

Fig. 5.1 CO_2 sorption in coal seams [5]

[5]. However, in field conditions, in-situ sorption capacities may be lower due to variation in coal seam porosity and permeability. Gas adsorption might cause the coal matrix to shrink or swell leads to reduce storage capacity. The field tests and simulation modeling can be integrated with geophysical logs to understand the real field conditions.

5.5 Conclusion

CO_2 sequestration and CH_4 extraction from unmineable coal seams have dual benefit: reducing greenhouse gas emissions and improving energy recovery through coalbed methane (CBM) production. The studied unmineable coal seams in the Singrauli coalfield show great potential for both CO_2 storage and methane extraction. With estimated maximum sorption capacities of 10,758 Mm^3 for CO_2 and 3215 Mm^3 for CH_4, these seams demonstrate the feasibility of deep, unmineable coal seams for geologic carbon storage and energy recovery. Although, the actual reservoir conditions may provide lower sorption capacities due to some factors such as coal seam porosity, permeability, and the effects of matrix shrinkage and swelling. Therefore, a detailed investigation with simulation modelling is essential for prediction of real field scale estimation. However, the unmineable coal seams are highly prospective for combining carbon sequestration and methane extraction. This opportunity can contribute to both the climate change mitigation and the development of cleaner energy sources.

References

1. IEA (2022) Global energy review: CO_2 emissions in 2021 global emissions rebound sharply to highest ever level international energy
2. CMPDI Annual Report (2022) Pli. In: Annual report of CMPDI on coal inventory
3. Das S, Dutta P (2017) Preliminary understanding of CO_2 sequestration and enhanced methane recovery in Raniganj Coalfield of India by reservoir simulation. Energy Procedia 114:4643–4657. https://doi.org/10.1016/j.egypro.2017.03.1267
4. DGH India (2019) Coal bed methane (CBM) I Directorate general of hydrocarbons (DGH). MoC India. http://www.dghindia.org/index.php/page?pageId=38
5. Singh S, Boruah A, Devaraju J (2023) CO_2 sequestration and CH_4 extraction from unmineable coal seams in Singrauli coalfield. Int J Coal Prep Utilization 1–18. MoS&PI India (2019) Ministry of statistics and program implementation I Government of India. http://www.mospi.gov.in/national-sample-survey-office-nsso
6. Dutta P, Harpalani S, Prusty B (2008) Modeling of CO_2 sorption on coal. Fuel 87(10–11):2023–2036. https://doi.org/10.1016/j.fuel.2007.12.015. Ojha K, Karmakar B, Mandal A, Pathak AK (2011) Coal bed methane in India: difficulties and prospects. Int J Chem Eng Appl January 2016, 256–260. https://doi.org/10.7763/ijcea.2011.v2.113
7. Hawkins JM, Houston R, Schraufnagel RA, Olszewski AJ, Pittsburgh R (1992) Estimating coalbed gas content and sorption isotherm using well log data. Soc Petrol Eng Inc. Pini R

(2009) Research collection. BRISK Binary Robust Invariant Scalable Keypoints 12–19. https://doi.org/10.3929/ethz-a-010782581

8. IEA (2019) Emissions—global energy & CO_2 status report 2019—analysis—IEA. Iea. https://www.iea.org/reports/global-energy-co2-status-report-2019. Rogers RE (1994) Coalbed-methane-principles-and-practices.pdf. In: Lake LW (ed) PTR Prentice Hall 113 Sylvan Aveenue Englewood Cliffs, NJ 07632
9. Asif M, Wang L, Panigrahi DC, Ojha K, Hazlett R (2022) Integrated assessment of CO_2-ECBM potential in Jharia Coalfield, India. Sci Rep 12(1):1–13. https://doi.org/10.1038/s41598-022-10574-5. Sakurovs R, Day S, Weir S (2012) Relationships between the sorption behaviour of methane, carbon dioxide, nitrogen and ethane on coals. Fuel 97:725–729. https://doi.org/10.1016/j.fuel.2012.03.014
10. Tang X, Ripepi N (2016) Temperature-dependent Langmuir's model in the coal and methane sorption process: statistics relationship. In: 2016 SME annual conference and expo: the future for mining in a data-driven world, January, 242–249. https://doi.org/10.19150/trans.7328
11. Zhang L, Aziz N, Ren T, Wang Z (2011) Influence of temperature on the gas content of coal and sorption modelling 269–276
12. Boruah A, Rasheed A, Mendhe VA, Ganapathi S (2019) Specific surface area and pore size distribution in gas shales of Raniganj Basin, India. J Petrol Explor Prod Technol 9:1041–1050
13. Hazarika S, Boruah A (2022) Supercritical CO_2 (SCO_2) as alternative to water for shale reservoir fracturing. Mater Today: Proc 50:1754–1757
14. Hazarika S, Boruah A, Saraf S (2023) Modeling and simulation study of CO_2 fracturing technique for shale gas productivity: a case study (India). Arab J Geosci 16(7):408
15. Boruah A, Prusty BK (2023) Influence of mineralogy, organic matter and thermal maturity on the diagenetic evolution of pore systems in the shales of Cambay and Raniganj Basin, India. J Geol Soc India 99(11):1577–1585

Chapter 6
CO_2 Sequestration in Saline Aquifers: Principles, Site Selection, and Operational Considerations

Abstract The chapter discusses the overview of CO_2 sequestration in salinc aquifers, and its significance as an important strategy to reduce global warming by minimizing CO_2 emissions. It explains fundamental principles like CO_2 phase behaviour, interactions with water and rock, and various trapping mechanisms associated with this process. The storage potential of saline aquifers is explained, focusing on key parameters such as storage capacity and CO_2 injectivity, which vastly influence their suitability for CO_2 storage. A detailed site selection process is also analysed, considering basin characteristics and reservoir parameters, as well as economic and social considerations.

Keywords CO_2 sequestration · Saline aquifers · Site selection · Operational considerations

6.1 Introduction

The rising concentration of carbon dioxide (CO_2) in the atmosphere, primarily influenced by the global consumption of fossil fuels such as oil, natural gas, and coal, has become a critical environmental concern. Over the last 250 years, CO_2 levels have been increasing from approximately 280 ppm to over 380 ppm, with an annual growth rate of about 1 ppm. This increase is directly affecting to measurable global warming, prompting the need for effective solutions to mitigate CO_2 emissions [1, 2]. As a result, CO_2 sequestration atracts attention as a significant strategy to mitigate the climate change by capturing and securely storing the CO_2 emissions. Among the various sequestration methods, saline aquifers represent one of the most promising options for the long-term CO_2 storage [1, 2]. Saline aquifers are the porous, permeable rock formations filled with highly saline water, and they have vast storage capacity. The saline acquifers have wide geographic distribution. The potential for CO_2 sequestration in these aquifers has been demonstrated in several field projects, with the Sleipner West Field in the North Sea being a significant example, where CO_2 has been injected into the Utsira Formation, a saline sandstone aquifer, since

A. Boruah, *CO_2 Geosequestration: Capturing Carbon for a Sustainable Future*, Springer Climate, https://doi.org/10.1007/978-3-031-81021-3_6

1996. Other notable sites, such as the Southern North Sea basin and the East Irish Sea basin in the United Kingdom, have also been identified as suitable candidates for CO_2 storage, based on specific geological criteria including lithology, porosity, depth, permeability, and salinity. For effective CO_2 storage, saline aquifers must meet several key criteria, such as sufficient depth to maintain CO_2 in a supercritical state, adequate porosity and permeability for fluid movement, and the presence of an effective cap rock to prevent CO_2 migration [1–3]. Additionally, storage sites must be a stable geological site which can reduce the risk of leakage due to tectonic activity. Understanding the behaviour of CO_2 under different pressure and temperature conditions, particularly its phase behaviour as a supercritical fluid, is important for determining storage efficiency and capacity [2–4]. This chapter provides a comprehensive understanding of the principles, factors, and of CO_2 sequestration mechanisms in saline aquifers, the selection of appropriate storage sites, and the operational considerations are necessary for successful CO_2 injection and containment. It explains the mechanisms of CO_2 trapping in confined and unconfined aquifers, assessing the storage potential of saline formations, and also discusses the economic sustainability of this method. Through a systematic approach to site selection and an understanding of CO_2 behaviour in subsurface environments, saline aquifers offer a promising solution for large-scale CO_2 sequestration, contributing to global efforts to reduce carbon emissions and mitigate climate change [2, 4, 5]. When carbon dioxide (CO_2) is injected into a saline aquifer, it dissolves into the formation water or brine, initiating a series of chemical interactions. Initially, CO_2 simply dissolves in the water, forming a solution. This leads to the establishment of an equilibrium between dissolved CO_2 and carbonic acid (H_2CO_3), which then dissociates into bicarbonate (HCO_3^-) and carbonate (CO_3^{2-}) ions. The carbonate anions can further react with cations such as calcium (Ca^{2+}) and magnesium (Mg^{2+}) present in the formation water, resulting in the precipitation of carbonate minerals like limestone and dolomite. This mineralization process has contributed to the formation of extensive carbonate rock deposits over geological time scales [1].

The interaction between CO_2 and brine also affects the CO_2 injection and flow dynamics within the aquifer. Firstly, when CO_2 dissolves in brine, it increases the density of the brine. Secondly, CO_2 reacts with water to form carbonic acid, which can lower the pH and make the water more acidic. Lastly, as water dissolves into the CO_2, the salinity of the remaining brine increases. The solubility of CO_2 in brine is influenced by several factors, with higher pressure increasing solubility and higher temperature and salinity reducing it.

6.2 CO_2 Injection into the Saline Aquifer

As CO_2 is injected into the saline aquifer of porous rock, it spreads and displaces the formation water, becoming trapped through a combination of physical and chemical mechanisms. These trapping mechanisms include geological trapping, where the

CO_2 is held by the structural features of the rock and due to residual saturation trapping, where CO_2 is immobilized in the pore spaces due to capillary pressure [4, 5]. Solubility trapping occurs when CO_2 dissolves into the formation water, and mineral trapping happens when CO_2 reacts with the rock matrix to form stable minerals. In addition, migration trapping involves CO_2 slowly moving through the subsurface at extremely low velocities, potentially remaining underground for millions of years [6]. The storage capacity of a saline aquifer is determined by factors such as pore volume, porosity, and the efficiency of both microscopic and volumetric displacement. These variables collectively determine the amount of CO_2 that can be securely stored in the formation, making saline aquifers a promising site for long-term CO_2 sequestration [6, 7]. The controlling factors are, density of CO_2 at reservoir conditions, the interconnected pore space volume, and fluid property [8–10]. Solubility affects storage capacity significantly because it determines the amount of gas that dissolves in the brine [11, 12]. The reactions among CO_2, brine, and the formation rock can change formation porosity and permeability (not affected by quartz, slightly affected by illite and kaolinite, and strongly affected by Calcite and Dolomite).

Injectivity is assessed by the reservoir quality, geometry, connectivity, and the interaction between CO_2, water, and rock [7–9]. The storage site should be located in a geologically stable area to get ride of the release of stored CO_2 due to tectonic movements (Fig. 6.1). Preferred sedimentary basins include intracratonic, foreland, and passive-margin basins, with deep, regionally extensive flow systems [10–15]. Cold basins are more suitable than warm basins due to their lower surface temperatures and geothermal gradients. Ideally, the basin should be rich in energy resources, with most hydrocarbon resources already extracted and advanced production methods in place. Reservoir Characteristics: Faults, fractures, and unconformities are considered unfavourable as they can create pathways for CO_2 to escape through the cap rock to the surface [16]. The cap rock, a low-porosity, impermeable layer, serves as the primary trapping mechanism, ensuring the long-term containment of CO_2 [14–20].

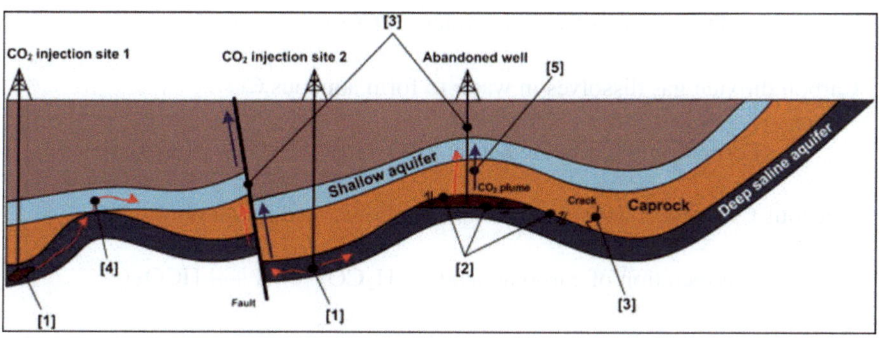

Fig. 6.1 CO_2 injection in aquifers [21]

6.3 CO_2 Saline Water Reaction

In the CO_2 sequestration process, CO_2 dissolves into the aquifer's pore fluid near the injection site, and over time, both free and dissolved CO_2 diffuse through the reservoir, eventually dispersing throughout the aquifer [21]. After hundreds of years, dissolved CO_2 becomes the dominant form of trapped CO_2 in the reservoir [21]. Several factors influence solubility trapping, including temperature, pressure, salinity, and the composition of the formation rock [21, 22]. CO_2 solubility in saline water increases with pressure and decreases with higher temperature and salinity. For example, Bando et al. [22] demonstrated a 20% reduction in CO_2 solubility when temperature increased from 80 to 200 °C at 10 MPa. Pruess et al. [23] stated that CO_2 solubility in brine increases by 1 kg/m^3 for each 1 MPa rise in pressure and decreases by 0.4 kg/m^3 for every 1 °C temperature drop within the 10–30 MPa pressure and 35–100 °C temperature ranges. Additionally, higher salinity in brine reduces CO_2 solubility [24, 25]. Furthermore, recent studies [20–25] have explored techniques to enhance CO_2 dissolution in brine, such as injecting brine or water atop the CO_2 plume or pumping brine from saline aquifers. Injecting water beneath the cap rock after CO_2 injection could boost CO_2 dissolution from 20 to 77% over 200 years, with further increases achievable with higher pumping rates [23, 24]. These techniques are energy-efficient and could reduce CO_2 leakage risks while increasing storage capacity, though their real-world application has yet to be thoroughly tested [25]. The physical properties of the aquifer, such as formation thickness, tilt angle, permeability anisotropy, and layering, also impact CO_2 solubility. For instance, thinner formations promote greater contact between CO_2 and brine during injection, though solubility decreases after injection due to limited interaction [24, 25]. Aquifer tilt angles and permeability anisotropy ratios also affect CO_2 migration and mixing, with increased layers promoting CO_2 retention and contact with brine [25]. Here are the chemical equations for the solubility trapping process in saline aquifers:

$$\text{Dissolution of } CO_2 \text{ in water}: \quad CO_2(g) \rightleftharpoons CO_2(aq)$$

Carbon dioxide gas dissolves in water to form aqueous CO_2.

$$\text{Formation of carbonic acid}: \quad CO_2(aq) + H_2O \rightleftharpoons H_2CO_3$$

Aqueous CO_2 reacts with water to form carbonic acid (H_2CO_3).

$$\text{Dissociation of carbonic acid}: \quad H_2CO_3 \rightleftharpoons H + +HCO_3$$

Carbonic acid partially dissociates into hydrogen ions (H^+) and bicarbonate ions (HCO_3^-).

$$\text{Further dissociation of bicarbonate}: \quad HCO_3 - \rightleftharpoons H + +CO_2^3$$

Bicarbonate ions further dissociate into hydrogen ions (H^+) and carbonate ions (CO_3^{2-}).

These reactions describe how CO_2 is dissolved and chemically interacts with water and other ions in saline aquifers, leading to long-term CO_2 storage through solubility trapping.

6.4 Mineral Trapping of CO_2 in Saline Aquifers

Mineral trapping refers to the process in which dissolved CO_2 reacts with formation rocks, leading to the long-term storage of CO_2 in saline aquifers. This slow process, taking hundreds to thousands of years, relies on the presence of reactive minerals like feldspars, clays, micas, and iron oxides. Over time, more than 90% of injected CO_2 can be permanently trapped as solid carbonates such as calcite ($CaCO_3$), magnesite ($MgCO_3$), dolomite ($MgCO_3 \cdot CaCO_3$), and siderite ($FeCO_3$). Although mineral trapping usually occurs in an aqueous environment, some reactions can happen without water, such as brine dissolution near the injection well where CO_2 concentrations are high [22–25]. When CO_2 is injected into a saline aquifer, two types of reactions can occur: pH-buffered and non-pH buffered. The pH of the aquifer can drop to around 3 near the injection site but gradually increase to 7 as reactions progress and minerals buffer the pH. These reactions promote solubility trapping by forming dissolved bicarbonate ions, sometimes referred to as ionic trapping [2]. Dissolving formation rock minerals increases porosity and permeability near the injection site, aiding CO_2 trapping. However, early-stage carbonate precipitation can reduce aquifer permeability, making the process complex and requiring careful understanding [22]. Sandstone, carbonate, shale, and basalt formations are considered suitable for CO_2 storage, with sandstone offering the highest potential for pH buffering and promoting efficient CO_2 trapping [22]. Table 6.1 provides an overview of various successful CO_2 injection projects in saline aquifers, highlighting their key parameters, status, and geochemical reactions.

6.5 Global Status

Several successful CO_2 storage projects (Fig. 6.2) based on saline aquifers demonstrate the feasibility of long-term CO_2 sequestration (Table 6.1). **Sleipner CO_2 Storage** in the North Sea, Norway, is one of the well-known projects which has been operational since 1996. In this project, CO_2 is injected at a depth of 800–1000 m into the Utsira Formation. The injection process is controlled by parameters such as pressure, temperature, and permeability. The high porosity of the Utsira Formation (35–40%) facilitates efficient CO_2 dissolution in brine and calcite precipitation for trapping of CO_2 [26, 27].

Table 6.1 CO_2 injection in saline aquifers worldwide [1, 22–32]

Project name	Location	Controlling parameters	Current status	Depth (m)	Formation	Storage capacity (Mt CO_2)	Porosity (%)
Sleipner CO_2 storage	North Sea, Norway	Pressure, temperature, permeability	Operational since 1996	800–1000	Utsira Formation	16–20	35–40
In Salah CO_2 Project	Algeria	Fault integrity, cap rock permeability	Suspended (2011)	1800	Krechba Formation	17	13–19
Quest CCS Project	Alberta, Canada	Reservoir pressure, temperature, and flow rate	Operational since 2015	2000	Basal Cambrian Sands	27	18–25
Illinois Basin – Decatur	Illinois, USA	CO_2 injection rate, rock-fluid interactions	Operational since 2011	2100	Mt. Simon Sandstone	5	Dec-18
Gorgon CO_2 Injection	Western Australia	Reservoir integrity, fault reactivation risk	Operational since 2019	2300	Dupuy Formation	120	25–30
Ketzin CO_2 Project	Germany	CO_2 injection rate, mineralogy of reservoir	Completed in 2013	630–700	Stuttgart Formation	2	20–26
Snøhvit CO_2 Storage	Barents Sea, Norway	Fault integrity, injection rate, cap rock	Operational since 2008	2500	Tubåen Formation	23	25

In Salah CO_2 Project in Algeria, operational from 2004 until its suspension in 2011, injected CO_2 into the Krechba Formation at 1800 m. This project is controlled by structural trapping system on fault integrity and cap rock permeability. The porosity of the formation is less (13–19%), CO_2 dissolutes in brine and interacts with clay minerals. However, the project was suspended due to some operational challenges [28].

The **Quest CCS Project** in Alberta, Canada project has been operational since 2015. This project injects CO_2 into the Basal Cambrian Sands at a depth of 2000 m. Key controlling factors include reservoir pressure, temperature, and flow rate. With a porosity of 18–25%, CO_2 dissolves in brine and bicarbonate ions form, facilitating long-term storage. The project has a storage capacity of 27 Mt CO_2, and it has a significant contribution in reducing carbon emissions [29].

In the USA, the **Illinois Basin—Decatur Project** began in 2011, injecting CO_2 into the Mt. Simon Sandstone at 2100 m. The project is controlled by CO_2 injection

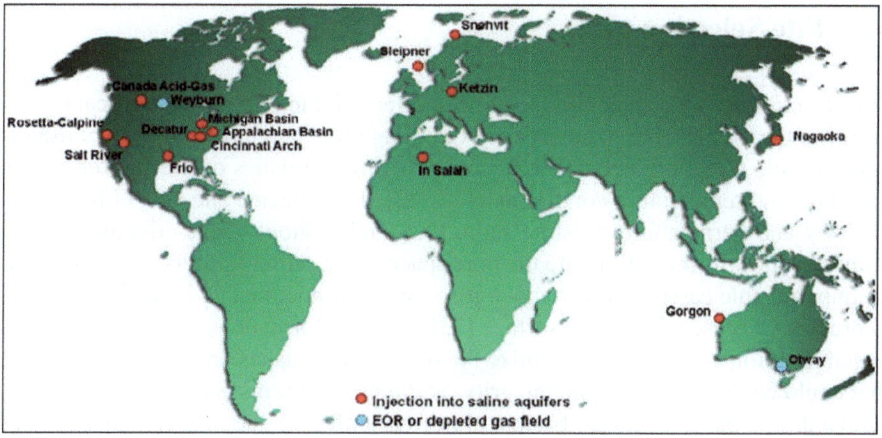

Fig. 6.2 Map showing CO_2 injection into deep saline aquifers (map is taken from open access journal) [11]

rates and rock-fluid interactions, with a porosity of 12–18%. Over time, CO_2 dissolves and forms solid precipitates such as calcite and siderite, providing effective storage [30].

Australia's **Gorgon CO_2 Injection Project** has been operational since 2019. Here the target formation is the Dupuy Formation at a depth of 2300 m. The main challenges of this project involve reservoir integrity and the risk of fault reactivation. With a porosity of 25–30%, it can store up to 120 Mt CO_2. Carbonate mineral precipitation and brine dissolution are the primary chemical reactions that secure the CO_2 in the formation [30].

Germany's **Ketzin CO_2 Project**, completed in 2013, and injected CO_2 into the Stuttgart Formation at a shallower depth of 630–700 m. Despite its relatively small capacity (2 Mt CO_2) and lower porosity (20–26%), the project demonstrated the formation of stable carbonate minerals such as dolomite and magnesite, providing insight into the effectiveness of mineral trapping [32].

Snøhvit CO_2 Storage project in the Barents Sea, Norway, has been operational since 2008. CO_2 is injected into the Tubåen Formation at a depth of 2500 m. The porosity of the formation is 25%, and the project is based on fault and cap rock integrity. The formation rocks underwent chemical reactions and CO_2 dissolved in brine. And the chemical reactions formed stable carbonates and bicarbonates to the long-term storage of 23 Mt CO_2 [31].

6.6 Site Selection

Site selection for CO_2 injection in saline aquifers (Table 6.2) plays a critical role in ensuring the effectiveness and security of storage for long term. The key considerations of site selection includes the geological characteristics of the aquifer, depth, pressure, and temperature conditions [1, 4, 22]. The selected site should have high porosity and permeability to store more CO_2 and it should have effective permeability to allow its migration into the pore spaces. The saline aquifers should possess an impermeable cap rock layer above it so that the impermeable rock acts as a seal to prevent CO_2 escaping to the surface [23]. Suitable formations are typically found in stable tectonic settings to minimize the risk of earthquakes or ground movements that could compromise storage integrity. Sedimentary basins, such as intracratonic, foreland, and passive-margin basins, are often preferred due to their large storage potential and geological stability [4, 5]. The depth of the aquifer is also critical, as deeper formations with higher pressure are more likely to maintain CO_2 in its dense supercritical state, enhancing storage efficiency [30].

The key considerations for operational process (Table 6.2) during CO_2 sequestration focus on managing the injection rate and process to ensure long-term stability and minimize environmental risks [23, 25, 34]. The rate of CO_2 injection monitoring is important to avoid excessive pressure build-up within the aquifer. This could lead to fracturing of the cap rock or fault reactivation, and potentially resulting in CO_2 leakage [21, 22]. Moreover, the injection process should be optimized to enhance CO_2 trapping mechanisms, such as solubility trapping, where CO_2 dissolves in brine, and mineral trapping, where it reacts with minerals to form stable carbonates. Monitoring systems must be appropriate and effective to track CO_2 flow in the reservoir. It is also important to ensure that it remains safely stored over time [24, 25]. The monitoring system includes use of seismic surveys, pressure sensors, and geochemical sampling to detect any alteration in the reservoir. Additionally, long-term modelling of reservoir behaviour is essential to estimate the CO_2 plume's migration and evaluate the potential for leakage [22].

6.7 Environmental Challenges

CO_2 injection in saline aquifers is an effective method for carbon sequestration. However, it comes with several environmental challenges:

Potential for Groundwater Contamination: CO_2 could migrate upward through faults, fractures, or improperly sealed wells, and can reach the overlying freshwater aquifers. This can result in contamination of groundwater resources, altering water chemistry and making it unsuitable for consumption. Dissolved CO_2 in formation water forms carbonic acid, which can lower pH levels and cause the dissolution of minerals like heavy metals (arsenic, lead) into groundwater, posing health risks.

Table 6.2 Critical factors to consider when selecting a saline aquifer for CO_2 sequestration and the operational measures necessary to ensure safe and effective CO_2 storage [4, 5, 15, 22, 30, 33]

Category	Criteria/Consideration	Description
Site selection criteria	Geological stability	The site must be in a tectonically stable region with minimal risk of earthquakes or ground movement that could compromise storage integrity
	Depth of aquifer	Ideal depth is greater than 800 m to maintain CO_2 in a dense supercritical state, enhancing storage efficiency
	Porosity and permeability	High porosity allows more CO_2 to be stored, while high permeability enables easy injection and migration of CO_2 into the formation
	Cap rock integrity	The aquifer must have an impermeable cap rock (e.g., shale) to prevent CO_2 from escaping to the surface, ensuring long-term containment
	Aquifer salinity	Saline aquifers (non-potable water) are preferred to avoid contamination of drinking water sources. High salinity reduces CO_2 solubility
	Reservoir size and storage capacity	The aquifer should have significant storage potential to accommodate large volumes of CO_2 over time
	Geological structure	The formation should be free of faults, fractures, and unconformities that could provide pathways for CO_2 leakage
	Reservoir composition	Aquifers with reactive minerals (e.g., feldspar, clay minerals) are preferred for mineral trapping, enhancing CO_2 immobilization
Operational considerations	Injection rate and pressure	The CO_2 injection rate must be controlled to prevent overpressure in the reservoir, which could lead to fracturing of the cap rock or fault reactivation
	Monitoring and verification	Continuous monitoring (seismic surveys, pressure sensors, and geochemical sampling) is required to track the CO_2 plume and ensure secure storage
	CO_2 trapping mechanisms	Solubility trapping (CO_2 dissolves in brine) and mineral trapping (CO_2 reacts with minerals to form stable carbonates) must be maximized to ensure long-term storage
	Well integrity	Injection wells must be designed and maintained to prevent leaks through casing or wellbore failure
	Cap rock and fault monitoring	Monitoring for cap rock integrity and potential fault reactivation is necessary to prevent unintended CO_2 migration
	Pressure management	Managing reservoir pressure is critical to avoid excessive build-up that could compromise storage integrity or lead to fault slippage

(continued)

Table 6.2 (continued)

Category	Criteria/Consideration	Description
	Brine displacement and management	Displacement of saline water by injected CO_2 must be carefully managed to avoid contamination of adjacent freshwater resources or unintended migration pathways
	Long-Term modelling and simulation	Simulation models should be used to predict the behaviour of CO_2 in the reservoir over time, including plume migration, trapping mechanisms, and pressure changes

Induced Seismicity: Injecting large volumes of CO_2 can increase subsurface pressure, potentially triggering small earthquakes, known as induced seismicity. These seismic events may cause fault and fractures that could provide pathways for CO_2 to escape. Pre-existing faults may be reactivated due to pressure changes, threatening the integrity of cap rocks designed to contain CO_2.
Caprock Integrity and Leakage: Over time, chemical reactions between CO_2, brine, and the rock matrix could weaken the caprock. This may reduce its sealing capacity and create pathways for CO_2 to escape back into the atmosphere or migrate to other parts of the subsurface. While mineral trapping is a long-term sequestration benefit, it can also reduce the permeability of the formation, potentially leading to blockages that alter the flow of CO_2 or increase pressure on the caprock.
Brine Displacement: The injection of CO_2 can displace brine, and can push it into freshwater aquifers or toward the surface. This can result in salinization of potable water resources, and can damage the ecosystems and agriculture.
Uncertainty in Long-Term Behaviour: Predicting the behaviour of injected CO_2 over centuries is complex due to the variability in geological formations. Long-term monitoring is essential, but ensuring the technology and infrastructure to maintain this over decades is an ongoing challenge. Current models may not fully capture the dynamic and multi-phase processes (such as solubility, mineral trapping, and CO_2 migration) in complex geological environments, leading to underestimation of potential risks.
Energy Use and Costs: Compressing CO_2 for injection and maintaining the necessary injection pressure requires significant energy, which may counteract some of the benefits of CO_2 sequestration until the energy source is renewable or low-carbon. For CO_2 leakage or other environmental impacts, the costs associated with mitigation, remediation, and ongoing monitoring are very high.
Impact on Ecosystems: The i CO_2 can changes water chemistry, which may impact subsurface microbial communities, altering the ecosystem services they provide, such as nutrient cycling and biodegradation.

References

1. Oelkers EH, Cole DR (2008) Carbon dioxide sequestration a solution to a global problem. Elements 4(5):305–310
2. Kabir M, Habiba UE, Khan W, Shah A, Rahim S, Patricio R, Shafiq M et al (2023) Climate change due to increasing concentration of carbon dioxide and its impacts on environment in 21st century; a mini review. J King Saud University-Sci 35(5):102693
3. White MT, Bianchi G, Chai L, Tassou SA, Sayma AI (2021) Review of supercritical CO_2 technologies and systems for power generation. Appl Therm Eng 185:116447
4. Crespi F, Gavagnin G, Sánchez D, Martínez GS (2017) Supercritical carbon dioxide cycles for power generation: a review. Appl Energy 195:152–183
5. Cole DR, Chialvo AA, Rother G, Vlcek L, Cummings PT (2010) Supercritical fluid behavior at nanoscale interfaces: implications for CO_2 sequestration in geologic formations. Phil Mag 90(17–18):2339–2363
6. Ringrose PS, Furre AK, Gilfillan SM, Krevor S, Landrø M, Leslie R, Zahid A et al (2021) Storage of carbon dioxide in saline aquifers: physicochemical processes, key constraints, and scale-up potential. Annu Rev Chem Biomol Eng 12(1):471–494
7. Luo A, Li Y, Chen X, Zhu Z, Peng Y (2022) Review of CO_2 sequestration mechanism in saline aquifers. Nat Gas Ind B 9(4):383–393
8. Izadpanahi A, Blunt MJ, Kumar N, Ali M, Tassinari CCG, Sampaio MA (2024) A review of carbon storage in saline aquifers: mechanisms, prerequisites, and key considerations. Fuel 369:131744
9. Mim RT, Negash BM, Jufar SR, Ali F (2023) Minireview on CO_2 storage in deep saline aquifers: methods, opportunities, challenges, and perspectives. Energy Fuels 37(23):18467–18484
10. White CM, Strazisar BR, Granite EJ, Hoffman JS, Pennline HW (2003) Separation and capture of CO_2 from large stationary sources and sequestration in geological formations—coalbeds and deep saline aquifers. J Air Waste Manag Assoc 53(6):645–715
11. Celia MA, Bachu S, Nordbotten JM, Bandilla KW (2015) Status of CO_2 storage in deep saline aquifers with emphasis on modeling approaches and practical simulations. Water Resour Res 51(9):6846–6892
12. Ren B, Hascakir B, Delshad M, Duncan IJ, Sepehrnoori K (2024) Carbon capture and storage at scale. Fuel 133512
13. Eyitayo S, Arbad N, Okere C, Gamadi T, Watson M (2024) Advancing geological storage of carbon dioxide (CO_2) with emerging technologies for climate change mitigation. Int J Environ Sci Technol 1–34
14. Li S, Wang P, Wang Z, Cheng H, Zhang K (2023) Strategy to enhance geological CO_2 storage capacity in saline aquifer. Geophys Res Lett 50(3), e2022GL101431
15. Wei YM, Kang JN, Liu LC, Li Q, Wang PT, Hou JJ, Yu B et al (2021) A proposed global layout of carbon capture and storage in line with a 2 C climate target. Nature Climate Change 11(2):112–118
16. Hesse MA, Tchelepi HA, Orr FM (2006, September). Scaling analysis of the migration of CO2 in saline aquifers. In SPE Annual Technical Conference and Exhibition? (pp SPE-102796). SPE
17. Shukla R, Ranjith P, Haque A, Choi X (2010) A review of studies on CO_2 sequestration and caprock integrity. Fuel 89(10):2651–2664
18. Fitts JP, Peters CA (2013) Caprock fracture dissolution and CO_2 leakage. Rev Miner Geochem 77(1):459–479
19. Albertz M, Stewart SA, Goteti R (2023) Perspectives on geologic carbon storage. Front Energy Res 10:1071735
20. Ozotta O, Ostadhassan M, Liu K, Liu B, Kolawole O, Hadavimoghaddam F (2021) Reassessment of CO_2 sequestration in tight reservoirs and associated formations. J Petrol Sci Eng 206:109071
21. Wei N, Li X, Wang Y, Dahowski RT, Davidson CL, Bromhal GS (2013) A preliminary sub-basin scale evaluation framework of site suitability for onshore aquifer-based CO_2 storage in China. Int J Greenhouse Gas Control 12:231–246

22. Muhammed NS, Haq B, Al Shehri D, Al-Ahmed A, Rahman MM, Zaman E (2022) A review on underground hydrogen storage: insight into geological sites, influencing factors and future outlook. Energy Rep 8:461–499
23. Matos CR, Carneiro JF, Silva PP (2019) Overview of large-scale underground energy storage technologies for integration of renewable energies and criteria for reservoir identification. J Energy Storage 21:241–258
24. Ringrose PS (2018) The CCS hub in Norway: some insights from 22 years of saline aquifer storage. Energy Procedia 146:166–172
25. Johnson JW, Nitao JJ, Knauss KG (2004) Reactive transport modelling of CO_2 storage in saline aquifers to elucidate fundamental processes, trapping mechanisms and sequestration partitioning. Geol Soc, London, Spec Publ 233(1):107–128
26. Chadwick RA, Zweigel P, Gregersen U, Kirby GA, Holloway S, Johannessen PN (2004) Geological reservoir characterization of a CO_2 storage site: the Utsira Sand, Sleipner, northern North Sea. Energy 29(9–10):1371–1381
27. Arts R, Eiken O, Chadwick A, Zweigel P, Van Der Meer B, Kirby G (2004) Seismic monitoring at the Sleipner underground CO_2 storage site (North Sea)
28. Fibbi G, Del Soldato M, Fanti R (2022) Review of the monitoring applications involved in the underground storage of natural gas and CO_2. Energies 2023(16):12
29. Bourne S, Crouch S, Smith M (2014) A risk-based framework for measurement, monitoring and verification of the quest CCS project, Alberta, Canada. Int J Greenhouse Gas Control 26:109–126
30. Greenberg SE (2021) An assessment of geologic carbon sequestration options in the Illinois basin: phase III (No. DOE-UIUC-42588). Univ. of Illinois, Champaign, IL, United State
31. Estublier A, Lackner AS (2009) Long-term simulation of the Snøhvit CO_2 storage. Energy Procedia 1(1):3221–3228
32. Streibel M, Finley RJ, Martens S, Greenberg S, Möller F, Liebscher A (2014) From pilot to demo scale—Comparing Ketzin results with the Illinois Basin-Decatur project. Energy Procedia 63:6323–6334
33. Wang Y, Vuik C, Hajibeygi H (2022) CO_2 Storage in deep saline aquifers: impacts of fractures on hydrodynamic trapping. Int J Greenhouse Gas Control 113:103552
34. Iglauer S (2011) Dissolution trapping of carbon dioxide in reservoir formation brine-a carbon storage mechanism. INTECH Open Access Publisher, London, UK, pp 233–262

Chapter 7
Basalt: A Viable Host Rock for CO_2 Sequestration and Mineralization

Abstract Basalt is considered as a promising host rock for CO_2 sequestration due to its composition with minerals cations such as Ca, Mg, and Fe. Basalt is composed of olivine, pyroxene, and plagioclase feldspar which react with dissolved CO_2 and form carbonate minerals, which can trap the injected CO_2 in secure solid forms. CO_2 sequestration in basalt involves various ways and techniques that capitalize on the reactivity of basaltic minerals to capture and securely store carbon dioxide. This chapter discusses about different approaches for CO_2 sequestration in basalt, elucidating the mineralization process. It also highlights its potential as an effective carbon storage media.

Keywords Basalt · CO_2 sequestration · Mineralization · Cabfix

7.1 Introduction

As global concerns on climate change is rising day by day, the search for effective methods to reduce greenhouse gas emissions is became crucial [1, 2]. Carbon capture and storage (CCS) is one of the promising way, where carbon dioxide (CO_2) emissions from industrial sectors are captured and stored underground geological media to prevent their release into the atmosphere [1, 2]. Among various geological storage medias, CO_2 storage in basalt rocks has emerged as an innovative and sustainable solution (Fig. 7.1). This chapter discusses the mechanisms, advantages, and issues of carbon sequestration in basalt. It explains a viable long-term storage for reducing atmospheric CO_2.

Basalt is a common volcanic rock and it has unique properties that make it favourable for CO_2 storage [3–5]. It composed of minerals like calcium, magnesium, and iron, and these minerals can chemically react with CO_2 to form stable carbonate minerals. This process is known as mineral carbonation [4–7]. Unlike other geological media, where CO_2 remains trapped in its gaseous or supercritical state, carbon sequestration in basalt is mainly the formation of solid minerals, and it stored the carbon dioxide for million years [1–4, 6–11].

A. Boruah, *CO_2 Geosequestration: Capturing Carbon for a Sustainable Future*, Springer Climate, https://doi.org/10.1007/978-3-031-81021-3_7

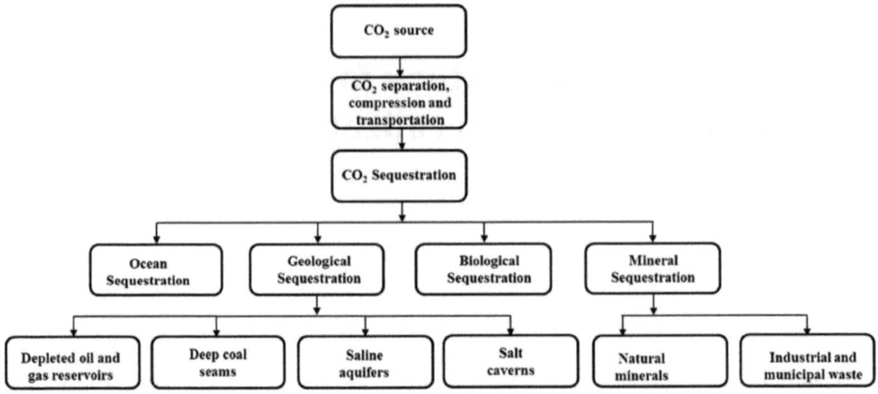

Fig. 7.1 Flow chart showing the options for CO_2 sequestration

The basalt is an abandoned rock which is present both on land and beneath the ocean floor. Its occurrence in large geographical areas presents a significant opportunity for large-scale carbon storage [3–5]. Basalt formations are occurring worldwide, and the major deposits are found in the Pacific Northwest of the United States, Iceland, and India, among others [3–5, 12, 13]. The Sleipner gas field, located west of Norway, is one of the classic example of CO_2 storage in basalt where approximately 1 million metric tons of CO_2 have been injected annually since 1996, and stored more than 700 m below the ocean floor in the Utsira sandstone formation [14–16]. This CCS site presents the world's longest-running carbon capture and storage (CCS) operation site [16, 17]. However, in this type of media, the transformation of CO_2 into stable carbonate mineral takes long time through a process known as *mineral trapping, which* takes tens of thousands of years. The primary drawbacks in such sedimentary formations is the low reactivity of silicate minerals, which slows down the mineralization process. To address these challenges, a promising alternative is found out where the CO_2 is injected into the basaltic rocks. This approach offers several key advantages. Moreover, basaltic rocks are abundant, covering approximately 10% of the continental landmass and much of the ocean floor [3, 4]. This widespread availability suggests that basalt carbonation could become a crucial component of global carbon storage strategies. As the basalt is more reactive than other rock types, it can highly react to do CO_2 mineralization, this method provides a more efficient ways of permanent carbon storage in comparison with other geological sites [18–20]. There is a challenge that persists in both sedimentary and basaltic rock storage is the buoyancy of CO_2. As CO_2 is less dense than the surrounding fluids, it has a natural tendency to rise towards the Earth's surface through pores and fractures in the rock formations [1]. Because of this buoyancy CO_2 can escape back into the atmosphere. To get ride of this issue, CO_2 can be injected in to the dissolved water. This ensures the integrity of the storage reservoir preventing the leakage and it also ensures long-term carbon sequestration [20]. The high opportunities of carbonation process in basalt and its abundance on Earth make it one of the most prospective candidate for overcoming the limitations of sedimentary rock storage, which offers a faster, and more secure method for mitigating atmospheric CO_2 levels [3, 4, 20–22].

The process behind CO_2 storage in basalt is mineral carbonation, a natural reaction where CO_2 dissolves in water and it reacts with the minerals in basalt to form carbonates [23, 24]. The overall reaction is as follows:

$$CO_2 + H_2O + Ca/Mg/Fe \rightarrow CaCO_3/MgCO_3/FeCO_3$$

When CO_2 is injected into basalt formations, it initially dissolves in water, forming carbonic acid (H_2CO_3). This acid reacts with the calcium, magnesium, and iron present in the basalt to form stable carbonate minerals such as calcite ($CaCO_3$), magnesite ($MgCO_3$), or siderite ($FeCO_3$). These minerals are solid, stable, and effectively permanent, This process mimics natural weathering, but occurs at an accelerated pace due to the controlled injection of CO_2 [25–28]. Studies have shown that in optimal conditions, significant carbonation can occur within a matter of years, much faster than originally anticipated.

7.2 CO_2 Storage in Basalt

Basalt is an extrusive igneous rock formed from the rapid cooling of basaltic lava at or near the Earth's surface [2–4]. It typically has a fine-grained texture due to the quick cooling process, which prevents the formation of large crystals. Basalt is primarily composed of mafic minerals, meaning it contains high levels of magnesium (Mg) and iron (Fe) and relatively low amounts of silica (SiO_2). This composition gives basalt its characteristic dark colour, ranging from black to dark gray. Structurally, basalt often occurs in large, layered flows that can span vast areas, especially in the case of continental flood basalts [3]. These formations can be hundreds of meters thick and are composed of multiple lava flow layers, making them ideal for geological studies and potential carbon storage.

Basalt, being rich in, offers an optimal environment for carbon mineralization. The minerals in basalt i.e. calcium (Ca), magnesium (Mg), and iron (Fe) silicates readily react with carbon dioxide (CO_2) to form stable carbonates like calcite ($CaCO_3$), magnesite ($MgCO_3$), and siderite ($FeCO_3$). The primary minerals in basalt are plagioclase, pyroxene, and olivine. Plagioclase feldspar, usually calcium-rich, is one of the most abundant minerals in basalt, followed by pyroxenes such as augite, which provide the rock with its rich iron and magnesium content. Olivine, another common mineral, further contributes to the mafic nature of basalt and its suitability for CO_2 mineralization [3, 22]. Basalt contains accessory minerals such as magnetite, ilmenite, and also minor amounts of quartz. The presence of these minerals makes basalt a highly favourable for carbonate reactions leading to mineralization and carbon storage [23, 24]. The recent studies demonstrated that the process is only controlled by the kinetics of the chemical reactions but also by the mixing behaviour of CO_2 in the subsurface [4, 5, 7, 23]. Appropriate mixing of CO_2 with water, basalt,

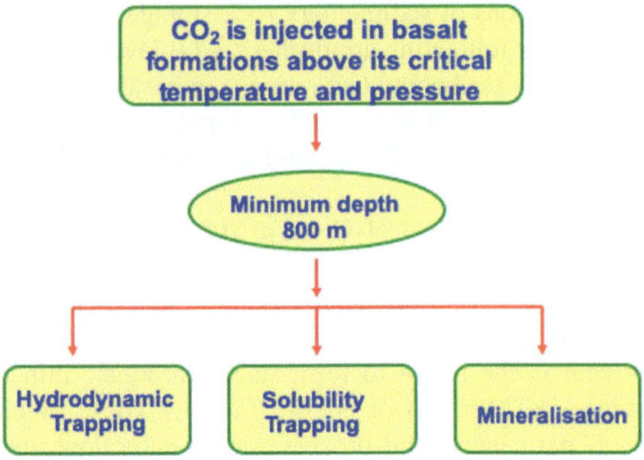

Fig. 7.2 CO_2 storage in basalt

and its minerals is crucial to enhance the rate of carbonation. The reservoir permeability, fluid flow dynamics, and the porosity of the basaltic rock are important to evaluate in determining how effectively and quickly CO_2 can be trapped in a mineralized state [25]. Therefore it is the important for proper site selection and reservoir management to ensure the success of carbon storage in basalt formations (Fig. 7.2).

7.3 Occurrences and Distribution of Basalt Rocks

Major continental basalt formations, known as large igneous provinces (LIPs), are found in regions like the Deccan Traps in India, the Columbia River Basalt Group in the United States, and the Siberian Traps in Russia [16, 17, 26, 27]. These regions are dominating by the extensive layers of basalt formed through volcanic activity, and during large-scale eruptions millions of years ago. Offshore basalt formations covers most of the ocean floor, and they are mainly located along mid-ocean ridges. In these regions, the tectonic plates diverge, and magma from the mantle rises to the surface, and creates new oceanic crust. Iceland has extensive basalt formations, straddles the Mid-Atlantic Ridge, the key location for basalt carbon storage sites [16–18, 26–28].

Different basalt formations around the world exhibits unique geological and mineralogical characteristics as they are formed in different geological time, underwent different geographical history and tectonic settings. For example, the Deccan Traps in India, formed around 65 million years ago during a period of intense volcanic activity, and it cover an area of approximately 500,000 square kilometres [16]. The basalt in this region is characterized by thick, layered flows, often several kilometres thick, with high magnesium content, which are suitable for mineral carbonation. Columbia River Basalt Group in the Pacific Northwest of the United States, formed over 17–6 million years ago, features thinner, more widespread basalt flows [26, 27]. This formation has columnar jointing, a characteristic pattern of vertical fractures that occur as the lava cools and contracts [17, 18, 26, 27]. The oceanic basalts, such

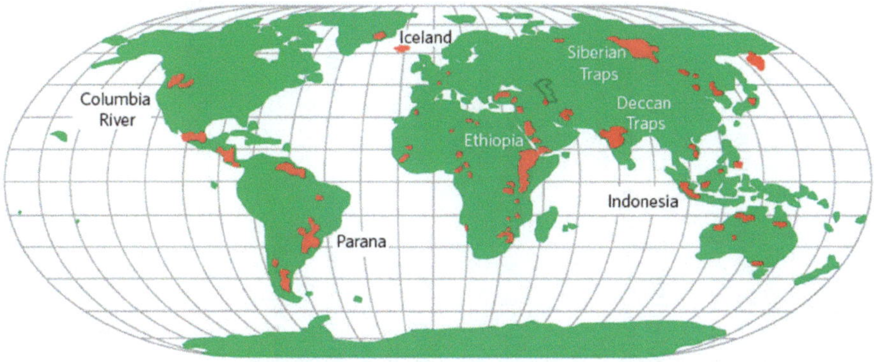

Fig. 7.3 Global distribution of basalt rocks [31]

as those found in Iceland and along mid-ocean ridges, are younger and typically have higher iron content than their continental basalts [26, 27]. The Mid-Atlantic Ridge, for instance, produces basalt that is rich in olivine and pyroxene, two key minerals that enhance its reactivity with CO_2. Iceland's basalt formations, which are continuously replenished by volcanic activity, have been the focus of several carbon storage initiatives, such as the CarbFix project [23]. These formations have demonstrated rapid CO_2 mineralization, highlighting the potential of basalt for large-scale carbon sequestration [23].

The basalt formations are widespread across the world, from the vast continental traps to the oceanic floors (Fig. 7.3). Their mineralogy, rich in calcium, magnesium, and iron oxides, combined with their high reactivity in water, makes them ideal candidates for carbon storage [23, 29, 30]. The variability in the physical and chemical characteristics of these basalt formations provides diverse opportunities for future CO_2 sequestration efforts, with each region offering unique advantages based on its geological history (Table 7.1).

7.4 Processes and Methods of Carbon Mineralization

Carbon mineralization, also known as *mineral carbonation*, is a natural process where carbon dioxide (CO_2) reacts with minerals, primarily silicates, to form stable carbonate minerals [4, 25, 26, 30]. This process locks away CO_2 in solid form, providing a secure and long-term solution for carbon sequestration. In nature, this reaction occurs slowly over geological time scales, but researchers have developed methods to accelerate it for carbon capture and storage (CCS) purposes. The key methods of carbon mineralization can be broadly categorized into in-situ and ex-situ approaches.

Table 7.1 A detailed table summarizing the characteristics, composition, and CO_2 storage potential of various basalt formations around the world

Sl. No.	Name	Country	Thickness	Age (M.Y.)	Depth	Mineral constituents	Chemical composition	Active CO_2 storage
1	Siberian Traps	Russia	Avg. 1–2 km; Max. 3–4 km	252	Varied (surface and subsurface)	Olivine ((Mg, Fe)$_2$SiO$_4$), Pyroxenes (e.g., augite, diopside), Plagioclase feldspar ((Ca, Na)AlSi$_3$O$_8$)	Silica (SiO_2): 45–55%, Al_2O_3, FeO, Fe_2O_3, CaO	No ongoing projects
2	Icelandic Basalts	Iceland	1–2 km	20	Near volcanic zones; shallow to 3–4 km	Olivine ((Mg, Fe)$_2$SiO$_4$), Pyroxenes (e.g., augite, diopside), Plagioclase feldspar ((Ca, Na)(Al, Si)$_4$O$_8$), Magnetite	Silica (SiO_2), Al_2O_3, Na_2O, K_2O	CarbFix—leading CO_2 storage project worldwide
3	Basalt of Indonesia	Indonesia	800–2000 m	May-23	Varied	Olivine, Pyroxene, Plagioclase feldspar, Magnetite	Ca, Mg, Fe, SiO_2	Still in developmental stages
4	Chilcotin Group	British Columbia	N/A	30	Near surface or shallow subsurface	Olivine, Plagioclase, Pyroxene	Rich in iron, magnesium, calcium	No ongoing projects
5	Karoo Basalt	South Africa	N/A	182–183	Varied; surface or beneath Karoo Supergroup	Plagioclase (Labradorite/ Bytownite), Pyroxene (Augite), Olivine (sometimes altered), Secondary minerals (zeolites, calcite)	Tholeiitic basalt (low SiO_2, high FeO and MgO)	Still in planning/ research phase

(continued)

Table 7.1 (continued)

Sl. No.	Name	Country	Thickness	Age (M.Y.)	Depth	Mineral constituents	Chemical composition	Active CO_2 storage
6	Paraná Traps	Brazil	A few tens of meters to 1.7 km	132–133	Varied	Plagioclase, Pyroxene (Augite), Olivine	Silicate minerals rich in Mg, Ca, Fe	No ongoing projects
7	Deccan Traps	India	>2000 m	66	Varied	Olivine, Pyroxene, Plagioclase, Basaltic glass	Tholeiitic basalts	Still in planning/ research phase
8	Rajmahal Trap Basalt	India	50–150 m	117–118	Shallow	Plagioclase (Anorthite-Albite), Pyroxenes, Augite, Olivine	SiO_2: ~48–50%, Al_2O_3: ~ 12–16%, Fe_2O_3: ~10–14%	No ongoing projects
9	Krishna-Godavari Basin	India	200–300 m	66	2000–3000 m	Plagioclase, Pyroxenes, Olivine	Various oxides, rich in Ca, Mg, Fe	No ongoing projects

7.4.1 In-Situ Carbon Mineralization

In-situ carbon mineralization involves injecting CO_2 directly into underground formations, such as basaltic rocks or ultramafic deposits, where the CO_2 reacts with minerals in the subsurface. Once injected, the CO_2 dissolves into formation water, forming weak carbonic acid (H_2CO_3). This acid reacts with the reactive minerals like olivine, pyroxene, and plagioclase, which are rich in calcium (Ca), magnesium (Mg), and iron (Fe). These reactions produce stable carbonate minerals like calcite ($CaCO_3$), magnesite ($MgCO_3$), and siderite ($FeCO_3$), permanently storing the CO_2 underground [5, 26, 31]. The success of in-situ mineralization process is controlled by many factors like the porosity and permeability of the rock formation, the availability of reactive minerals, and the rate at which CO_2 and water mix within the subsurface. CarbFix projects in Iceland have demonstrated the viability of in-situ carbon mineralization, where injected CO_2 mineralizes into solid carbonates within just a few years [23, 26]. In such projects, CO_2 is often injected dissolved in water to enhance mixing and reaction rates, significantly accelerating the carbonation process.

7.4.2 Ex-Situ Carbon Mineralization

Ex-situ carbon mineralization can be done on the ground, where CO_2 is reacted with crushed rocks or industrial byproducts in a controlled environment. This method involves mining and processing rocks which are rich in calcium and magnesium silicates, such as basalt, serpentinite, or olivine. The rocks are finely ground to increase their surface area, and then mixed with CO_2 under specific pressure and temperature conditions. The CO_2 reacts with the mineral surface to form carbonates, which can be safely stored or used as building materials [23–25]. Ex-situ methods can be more controlled and faster than in-situ mineralization because temperature, pressure, and mixing can be optimized to enhance reaction rates. However, this approach is energy-intensive due to the need for mining, crushing, and transporting rocks. To mitigate these challenges, researchers are exploring the use of industrial byproducts such as steel slag, coal fly ash, or mine tailings, which are already rich in calcium and magnesium oxides and require less processing. These waste materials, when reacted with CO_2, can form carbonates, potentially turning an environmental liability into a solution for carbon sequestration [26].

7.4.3 Carbon Mineralization Reactions

Several strategies are being explored to enhance both in-situ and ex-situ mineralization rates. One method is increasing the surface area of reactive minerals through mechanical grinding. This exposes more reactive sites to CO_2. In some cases,

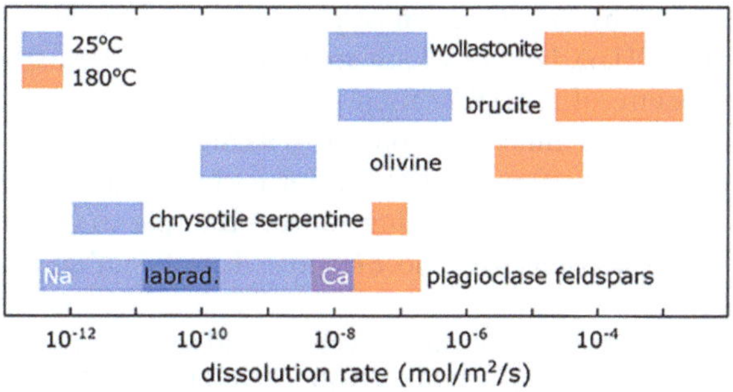

Fig. 7.4 Dissolution rates of magnesium- and calcium-rich minerals at 25 °C (blue) and 180 °C (orange)

increasing temperature and pressure can also speed up the reaction kinetics, though this requires more energy input. Another promising approach involves the use of additives or catalysts to accelerate the carbonation process. For example, the introduction of certain salts or chemicals can improve the dissolution of minerals and enhance CO_2 absorption [27–30]. Moreover, the biological methods are also being under research stage, where microbes produce carbonate minerals. These are helpful for facilitating the mineralization process. This approach, known as bio-mediated mineralization. This method offer a low-energy alternative for carbon capture, though it is still in the experimental stage [29].

The mineral wollastonite ($CaSiO_3$) exhibits the highest reaction rate (Fig. 7.4) with CO_2, ranging between 8.0×10^{-9} and 2.0×10^{-7} mol·m^{-2}·s^{-1} at 25 °C and 1.6×10^{-5} to 5.0×10^{-4} mol·m^{-2}·s^{-1} at 180 °C [4, 26]. However, its global availability is limited, with estimated reserves of only 100 million tons. Due to its high calcium content, wollastonite serves as a useful model for industrial byproducts like cement kiln dust, steel slag, and fly ash, which may exhibit similar initial reaction rates. However, these byproducts generally contain lower amounts of calcium or include calcium-aluminum silicates that are slower to react [25, 26].

Olivine (Mg_2SiO_4) reacts with CO_2-bearing fluids at a rate nearly comparable to wollastonite (Palandri & Kharaka, 2004; National Academies of Sciences, Engineering, and Medicine, 2019). As a key component of peridotite, which makes up much of Earth's upper mantle, olivine is highly abundant, making it a strong candidate for carbon mineralization. In single-step carbon mineralization experiments, adding $NaHCO_3$ buffers the pH, facilitating both dissolution and mineralization reactions [25–27]. Plagioclase feldspars, major components of basalt and gabbro in the oceanic crust, have slower reaction rates than wollastonite and olivine. At lower temperatures (25 °C), the calcium-rich anorthite ($CaAl_2Si_2O_6$) dissolves more rapidly than the sodium-rich albite ($NaAlSi_3O_6$). In basalts, plagioclase feldspar is often labradorite (50–70% anorthite), with reaction rates between 1.3×10^{-11} and 2.0×10^{-1} mol·m^{-2}·s^{-1} at 25 °C, and between 6.3×10^{-9} and 6.3×10^{-8} mol·m^{-2}·s^{-1} at 180 °C [3, 4, 18]. For industrial waste products containing Ca-Al silicates similar to anorthite, calcite ($CaCO_3$) formation may be hindered due to the slow diffusion of aluminum. Serpentine ($Mg_3Si_2O_5(OH)_4$) dissolves at rates comparable to the lower end of the range for plagioclase feldspars [3, 5, 19]. Serpentinite, formed by the alteration of peridotite, is often associated with olivine and brucite ($Mg(OH)_2$), both of which are highly reactive in carbon mineralization. The fibrous structure of asbestiform chrysotile, with its nano-scale cylindrical shape, accelerates carbon mineralization due to its high surface area-to-volume ratio. Additionally, using asbestiform serpentine for carbon mineralization reduces health risks from asbestos present in mining waste. Amorphous basaltic glass and certain industrial byproducts tend to be more reactive than crystalline minerals of the same composition [3].

During carbonation, a layer of solid reaction products may form on the surface of minerals, reducing their reactivity and limiting the extent of the reaction (a process known as "passivation"). However, in most studies, olivine does not show significant passivation, and the rate of carbonation remains steady, likely due to reaction-induced cracking and the formation of etch pits. In contrast, other minerals may experience reduced reaction rates over time, which could be problematic for extensive carbonation of basalts rich in plagioclase and pyroxenes. Understanding the factors that promote reaction-driven cracking, rather than surface passivation, is crucial for optimizing carbon mineralization processes [14, 15, 19, 26].

The estimated total storage capacities for in-situ CO_2 sequestration range from 8.0×10^3 to 3.5×10^4 $GtCO_2$ in sedimentary formations, 1.0×10^3 to 2.5×10^5 $GtCO_2$ for onshore basalts, 8.0×10^2 to 1.0×10^5 $GtCO_2$ for submarine basalts, and 6.0×10^4 to 6.0×10^5 $GtCO_2$ in onshore peridotite aquifers (National Academies of Sciences, Engineering, and Medicine, 2019). Additionally, fractured peridotite aquifers beneath the seafloor may offer an additional 6.0×10^7 $GtCO_2$ of storage capacity. Given that current global CO_2 emissions are approximately 40 $GtCO_2$ per year, storage capacity is not a limiting factor for CO_2 mitigation. However, the maximum injectivity rate for CO_2 in underground reservoirs poses a challenge. In the U.S., injectivity rates for sedimentary reservoirs range from 0.001 to 60 $MtCO_2$ per year. These rates align with the necessary climate goals, which require the storage of 125 $GtCO_2$ by 2100 (National Academies of Sciences, Engineering, and Medicine, 2019).

7.5 Challenges and Opportunities

The primary challenge of carbon mineralization is that it exhibits slow natural reaction rate, particularly in low-reactivity minerals and rock formations. This requires increasing the reactivity of the mineral surfaces or improving the efficiency of CO_2 mixing with reactive minerals. The energy and cost required for processes are high which includes rock mining, transportation, and grinding also pose barriers to widespread implementation of ex-situ methods. Although, the abundance of basalt and other reactive rocks and the long-term stability of carbonates, makes mineralization one of the most promising and secure forms of carbon storage. Carbon mineralization presents opportunities beyond just CO_2 storage. The carbonates produced can be used in different ways including construction materials such as cement and aggregates, offering a potential economic incentive for companies to adopt this technology.

7.6 CarbFix Project

As the basalt high porosity and permeability significantly enhance its reactivity with CO_2, making it an ideal medium for carbon capture and storage. Wallula in Washington state, USA, and CarbFix in Iceland have demonstrated the basalt as one of the most potential rock for sequestration. Both the projects evaluated the reservoir's composition, structure, and hydrology to assess their suitability for CO_2 injection and sequestration (National Academies of Sciences, Engineering, and Medicine, 2019). In 2013, the Wallula project injected 977 tons of CO_2 into basalt formations at pure state at depths from 838 to 886 m for over 25 days [26, 30].

The CarbFix project using CO_2 and H_2S from natural steam condensation to store. In Phase I, around 200 tons of CO_2 were injected at a depth of 500 m, with a 100-m flow path through basalt. They developed new methods during Phase I and Phase II, and separates water and CO_2 injection. It also ensures complete dissolution of CO_2 in water at that particular depth [26, 30]. This method does not require impermeable caprock, hence greatly expanding the scope of potential basalt storage reservoirs. Tracers like SF6 and 14C can be used to monitor CO_2 behaviour within the basalt aquifer, and rapid carbon mineralization removed dissolved CO_2 from the water within months. In Phase II, ongoing injections of 10–20 kilotons of CO_2 annually, along with a similar mass of H_2S, occur at depths of 1500 m, with an anticipated 2000-m flow path. The carbon mineralization has progressed at a rate of approximately 5×10^{-9} mol·m^{-2}·s^{-1} (comparable to experimental data shown in Fig. 7.2) in last three years. This approach also recycles pore water for continued co-injection with CO_2, addressing concerns over water usage in solution trapping techniques. Additionally, Climeworks has installed a direct air capture unit at the site to explore carbon-negative possibilities [23, 24, 30]. These projects underscore the significant potential for CO_2 storage in basalt formations.

7.7 Conclusion

CO_2 storage in basalt formations offers a promising method for long-term carbon sequestration. The geochemical characteristics of basalt, including its high reactivity with CO_2 due to minerals like olivine and plagioclase, make it an efficient medium for transforming CO_2 into stable carbonate minerals. Additionally, the natural porosity and permeability of basalt enhance its suitability for carbon capture, while its widespread availability on both land and in the oceanic crust provides vast potential for large-scale storage. Demonstrations such as the Wallula project in the USA and CarbFix in Iceland have confirmed the practicality of using basalt for

CO_2 storage. These projects have shown rapid mineralization rates and introduced innovative CO_2 injection techniques, proving basalt's ability to store carbon without requiring an impermeable caprock, thus broadening the range of possible storage sites. While basalt storage may involve higher costs compared to traditional sedimentary reservoirs, its large capacity and effectiveness, especially in volcanic and coastal regions, make it a valuable option for reducing global carbon emissions.

References

1. Hazarika S, Boruah A (2022) Supercritical CO_2 (SCO_2) as alternative to water for shale reservoir fracturing. Mater Today: Proc 50:1754–1757
2. Hazarika S, Boruah A, Saraf S (2023) Modeling and simulation study of CO_2 fracturing technique for shale gas productivity: a case study (India). Arab J Geosci 16(7):408
3. Irvine TN, Baragar WRAF (1971) A guide to the chemical classification of the common volcanic rocks. Can J Earth Sci 8(5):523–548
4. Middlemost EA (1989) Iron oxidation ratios, norms and the classification of volcanic rocks. Chem Geol 77(1):19–26
5. Bas ML, Maitre RL, Streckeisen A, Zanettin B, IUGS Subcommission on the Systematics of Igneous Rocks (1986) A chemical classification of volcanic rocks based on the total alkali-silica diagram. J Petrol 27(3):745–750
6. Walker GP (1993) Basaltic-volcano systems. Geol Soc, London, Spec Publ 76(1):3–38
7. Gislason SR, Broecker WS, Gunnlaugsson E, Snæbjörnsdóttir S, Mesfin KG, Alfredsson HA, Oelkers EH et al (2014) Rapid solubility and mineral storage of CO_2 in basalt. Energy Procedia 63:4561–4574
8. Xiong W, Wells RK, Horner JA, Schaef HT, Skemer PA, Giammar DE (2018) CO_2 mineral sequestration in naturally porous basalt. Environ Sci Technol Lett 5(3):142–147
9. Goldberg DS, Takahashi T, Slagle AL (2008) Carbon dioxide sequestration in deep-sea basalt. Proc Natl Acad Sci 105(29):9920–9925
10. Iglauer S, Al-Yaseri AZ, Wolff-Boenisch D (2020) Basalt-CO_2-brine wettability at storage conditions in basaltic formations. Int J Greenhouse Gas Control 102:103148
11. Al-Yaseri A, Ali M, Abbasi GR, Abid HR, Jha NK (2021) Enhancing CO_2 storage capacity and containment security of basaltic formation using silica nanofluids. Int J Greenhouse Gas Control 112:103516
12. Iglauer S, Al-Yaseri A (2021) Improving basalt wettability to de-risk CO_2 geo-storage in basaltic formations. Adv Geo-Energy Res 5(3):347–350
13. Wang Y, Chen Y, Wang W, Jiang F, Dou W, Su M, Bao Z (2024) Experimental research system and efficiency evaluation method of CO_2 sequestration in basalt mineralization. Energy Fuels 38(15):14414–14421
14. Carcione JM, Qadrouh AN, Alajmi M, Alqahtani NB, Ba J (2023) Rock acoustics of CO_2 storage in basalt. Geophys J Int 234(3):2429–2435
15. Snæbjörnsdóttir SÓ, Gislason SR (2016) CO_2 storage potential of basaltic rocks offshore Iceland. Energy Procedia 86:371–380
16. Liu D, Agarwal R, Liu F, Yang S, Li Y (2022) Modeling and assessment of CO_2 geological storage in the Eastern Deccan Basalt of India. Environ Sci Pollut Res 29(56):85465–85481
17. Rasool MH, Ahmad M (2023) Reactivity of basaltic minerals for CO_2 sequestration via in situ mineralization: a review. Minerals 13(9):1154
18. Gao Z, Xia C, Liao S, Yu X, Liu M, Li P, Liang X, Dai Q, Huang X (2023) Progress of methods for assessing CO_2 mineralization storage potential in basalt. Geol J China Universities 29(1):66
19. Matter JM, Broecker WS, Gislason SR, Gunnlaugsson E, Oelkers EH, Stute M, Sigurdardóttir H et al (2011) The CarbFix Pilot Project–storing carbon dioxide in basalt. Energy Procedia 4:5579–5585

20. Mohammed I, Yaseri A, Al Shehri D, Mahmoud M (2024) Basalt mineral surface charge and the effect of mineralization on its colloidal stability: implications of subsurface CO_2 storage. Fuel 356:129569
21. Anthonsen KL, Aagaard P, Bergmo PES, Erlström M, Fareide JI, Gislason SR, Snæbjörnsdottir SÓ et al (2013) CO_2 storage potential in the Nordic region. Energy Procedia 37:5080–5092
22. Kim K, Kim D, Na Y, Song Y, Wang J (2023) A review of carbon mineralization mechanism during geological CO_2 storage. Heliyon
23. Gislason SR, Wolff-Boenisch D, Stefansson A, Oelkers EH, Gunnlaugsson E, Sigurdardottir H, Fridriksson T et al (2010) Mineral sequestration of carbon dioxide in basalt: a pre-injection overview of the CarbFix project. Int J Greenhouse Gas Control 4(3):537–545
24. Zhao R, Zhong Z, Yu Y, Lü R, Shi T, Wang N, Cheng J (2024) Numerical simulation of CO_2 storage by basalts in Xingouzui formation, Jianghan Basin, China. Int J Greenhouse Gas Control 134:104133
25. Okoli AE, Kolawole O, Akaolisa CZ, Ikoro DO, Ozotta O (2024) Alterations in petrophysical and mechanical properties due to basaltic rock-CO_2 interactions: comprehensive review. Arab J Geosci 17(1):10
26. McGrail BP, Schaef HT, Ho AM, Chien YJ, Dooley JJ, Davidson CL (2006) Potential for carbon dioxide sequestration in flood basalts. J Geophys Res: Solid Earth 111(B12)
27. White SK, Spane FA, Schaef HT, Miller QR, White MD, Horner JA, McGrail BP (2020) Quantification of CO_2 mineralization at the Wallula basalt pilot project. Environ Sci Technol 54(22):14609–14616
28. McGrail BP, Spane FA, Amonette JE, Thompson CR, Brown CF (2014) Injection and monitoring at the Wallula basalt pilot project. Energy Procedia 63:2939–2948
29. Hosseini M, Ali M, Fahimpour J, Keshavarz A, Iglauer S (2022) Basalt-H2-brine wettability at geo-storage conditions: implication for hydrogen storage in basaltic formations. J Energy Storage 52:104745
30. Kelemen P, Benson SM, Pilorgé H, Psarras P, Wilcox J (2019) An overview of the status and challenges of CO_2 storage in minerals and geological formations. Front Clim 1:482595
31. Oelkers EH, Gislason SR, Matter J (2008) Mineral carbonation of carbon dioxide. Elements 4(333):337

Correction to: Reservoir Screening Criteria for CO_2 Storage

Correction to:
Chapter 2 in: A. Boruah, CO_2 Geosequestration: *Capturing Carbon for a Sustainable Future*, Springer Climate, https://doi.org/10.1007/978-3-031-81021-3_2

The original version of the book was inadvertently published without the word "carbon dioxide" and incorrectly mentioned as "carbon monoxide" in the chapter "Reservoir Screening Criteria for CO_2 Storage". All corrections listed below have been updated within the current version.

Corrections:

- **Chapter 2:**

  Page 20, Line 7, 9, 13 and 15 from the top
  Page 21, Line 8 and 17 from the bottom
  Page 22, Line 3, 8, 9, 10, 11, 13, 19, 21 and 27 from the bottom
  Page 23, Line 9, 18, 25, 27, 28, 40 and 41 from the top
  Page 24, Line 10, 11, 14, 15, 26, 28, 36 and 37 from the top
  Page 25, Line 4, 8, 17, 26, 32 and 39 from the top
  Page 26, Line 6, 8, 14, 16 and 29 from the bottom
  Page 29, Line 9 and 33 from the top
  Page 30, Line 6 and 8 from the bottom
  Page 32, Line 4 from the top
  Page 34, Line 3, 6 and 7 from the top, line 2 from the bottom

---

The updated version of this chapter can be found at
https://doi.org/10.1007/978-3-031-81021-3_2

A. Boruah, *CO_2 Geosequestration: Capturing Carbon for a Sustainable Future*,
Springer Climate, https://doi.org/10.1007/978-3-031-81021-3_8

MIX
Papier aus verantwortungsvollen Quellen
Paper from responsible sources
FSC® C105338

If you have any concerns about our products,
you can contact us on
ProductSafety@springernature.com

In case Publisher is established outside the EU,
the EU authorized representative is:
Springer Nature Customer Service Center GmbH
Europaplatz 3, 69115 Heidelberg, Germany

Printed by Libri Plureos GmbH
in Hamburg, Germany